FROM GLOBULAR PROTEINS TO AMYLOIDS

FROM GLOBULAR PROTEINS TO AMYLOIDS

Edited by

IRENA ROTERMAN-KONIECZNA

Department of Bioinformatics and Telemedicine, Jagiellonian University Medical College, Krakow, Poland

ELSEVIER

Elsevier
Radarweg 29, PO Box 211, 1000 AE Amsterdam, Netherlands
The Boulevard, Langford Lane, Kidlington, Oxford OX5 1GB, United Kingdom
50 Hampshire Street, 5th Floor, Cambridge, MA 02139, United States

Library of Congress Cataloging-in-Publication Data
A catalog record for this book is available from the Library of Congress

British Library Cataloguing-in-Publication Data
A catalogue record for this book is available from the British Library

ISBN: 978-0-08-102981-7

Publisher: Andre Gerhard Wolff
Acquisition Editor: Glyn Jones
Editorial Project Manager: Ali Afzal-Khan
Production Project Manager: Paul Prasad Chandramohan
Cover Designer: Matthew Limbert

Typeset by TNQ Technologies

Contents

Contributors

Mateusz Banach
Department of Bioinformatics and Telemedicine, Jagiellonian University Medical—College, Krakow, Poland

Dawid Dułak
Department of Biophysics, Faculty of Physics, Astronomy and Applied Computer Science — Jagiellonian University, Kraków, Poland

Piotr Fabian
Department of Theory of Informatics, Institute of Informatics, Silesian University of Technology, Gliwice, Poland

Małgorzata Gadzała
Cyfronet AGH Academic Computer Centre CYFRONET - University of Science and Technology in Cracow, Kraków, Poland; Currently - Schibsted Tech Polska Sp. z o. o., Kraków, Poland

Leszek Konieczny
Chair of Medical Biochemistry, Jagiellonian University—Medical College, Krakow, Poland

Irena Roterman
Department of Bioinformatics and Telemedicine, Jagiellonian University—Medical College, Krakow, Poland

Katarzyna Stapor
Department of Theory of Informatics, Institute of Informatics, Silesian University of Technology, Gliwice, Poland

Preface

The subject matter of this publication is the relationship between the aqueous solvent and the protein. We address numerous facets of the problem, including the effect exerted by water upon the polypeptide chain during folding, as well as changes which occur in the solvent due to the presence of dissolved proteins. Our model was initially formulated in 2006 [1], developed in Ref. [2] and is continually being refined.

The idea of recreating the spontaneous nature of folding approximating experimental conditions as much as possible with critical role of water environment was the basis for model development. Active participation of water environment produces a concentration of hydrophobicity at the center of the molecule, with values decreasing along with distance from the center to reach nearly 0 on the surface. These ideas introduced by Leszek Konieczny are based on long experience in biochemistry and medicine. Irena Roterman developed these ideas:

1. mathematically modeling the theoretical distribution of hydrophobicity with a 3D Gaussian,
2. applying Kullback-Leibler's divergence entropy formula to assess the differences between the theoretical (idealized) and observed distribution of hydrophobicity for a given protein,
3. applying concepts derived from information theory to characterize the variability of proteins, domains and other structures
4. identification and quantitative assessment (evaluation) of the local deformation, which appeared as carrying biological specificity of proteins
5. identifying the constitutive properties of amyloids as well as factors capable of arresting the propagation of fibrillary forms.

The former PhD student Mateusz Banach prepared the software capable of calculating T (theoretical), O (observed) and H (based on intrinsic hydrophobicity of each amino acid) distributions including the method of orienting the protein molecule in a coordinate system so that it can be embedded in a suitable 3D Gaussian capsule. Mateusz Banach also authored the vast majority of 3D figures and diagrams included in this publication.

Important contributions came from two other current postgraduate students — Dawid Dułak and Małgorzata Tomanek-Gadzała, who supplied polypeptide folding software applied in the section devoted to folding proteins and amyloids.

Katarzyna Stąpor and Piotr Fabian from Silesian University of Technology participated in folding simulation.

Other contributors of note include former PhD students — Dr Michał Bryliński (currently a professor at Louisiana State University) and Dr Wiktor Jurkowski. The authors are furthermore indebted to Piotr Nowakowski, who translated the work into English. Special thanks for critical comments which appeared creative.

Art-graphics was performed by Romuald Bolesławski.

Technical support provided by Anna Zaremba-Śmietańska and valuable remarks by Zdzisław Wiśniowski played a major part in the preparation of this work. The authors also acknowledge Academic Computing Center ACK Cyfronet AGH (University of Science and Technology), where some of the presented computations were carried out using the platform organized by PL Grid project [3].

Program PyMol was used for all 3D presentations [4]. and Matplotlib for profiles presentations [5].

Over the years this research was supported by the Jagiellonian University Collegium Medicum grant system, including the most recent grant (2016–9) K/ZDS/006363.

Irena Roterman-Konieczna
Kraków, Jan 2019.

REFERENCES

[1] Konieczny L, Brylinski M, Roterman I. Gauss-function-Based model of hydrophobicity density in proteins. In Silico Biology 2006;6(1–2):15–22.
[2] Roterman-Konieczna I. Protein Folding in Silico. 2012. https://doi.org/10.1533/9781908818256.
[3] http://plgrid.pl/.
[4] https://sourceforge.net/projects/pymol/The PyMOL Molecular Graphics System, version 1.8, Schrödinger LLC.
[5] Hunter JD. Matplotlib: a 2D graphics environment. Computing in Science & Engineering 2007;9(3):90–5. https://doi.org/10.1109/mcse.2007.55.

For interpretation of the references to colour in the figure legends, the reader is referred to the web version of this book.

Introduction

Leszek Konieczny, Irena Roterman

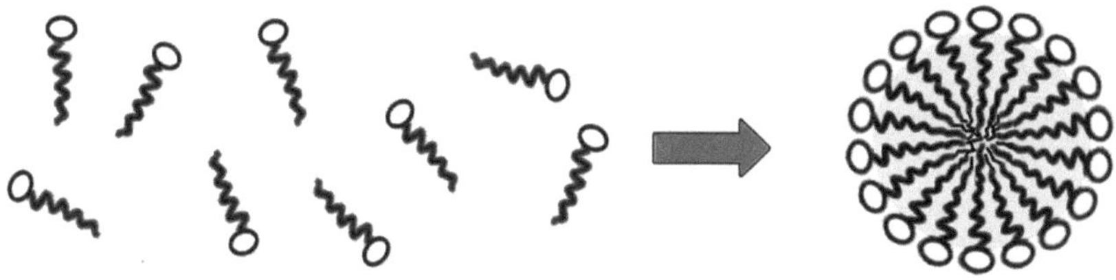

The diagram illustrates the basic idea behind the formation of spherical micelle constructed by identical bi-polar surfactant molecules.

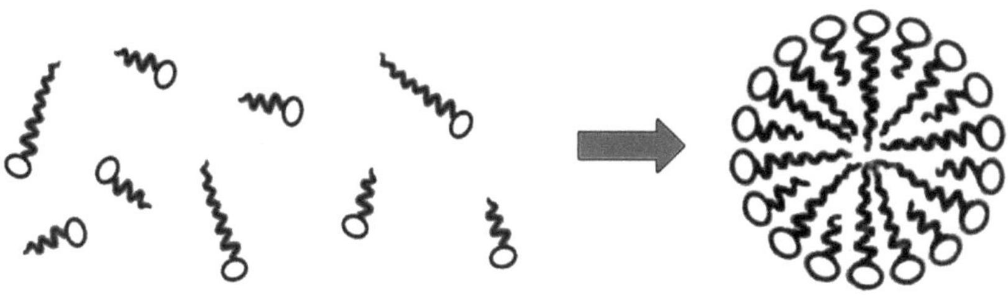

The diagram illustrates the basic idea behind the formation of spherical co-micelle constructed by two different forms of bi-polar surfactant molecules.

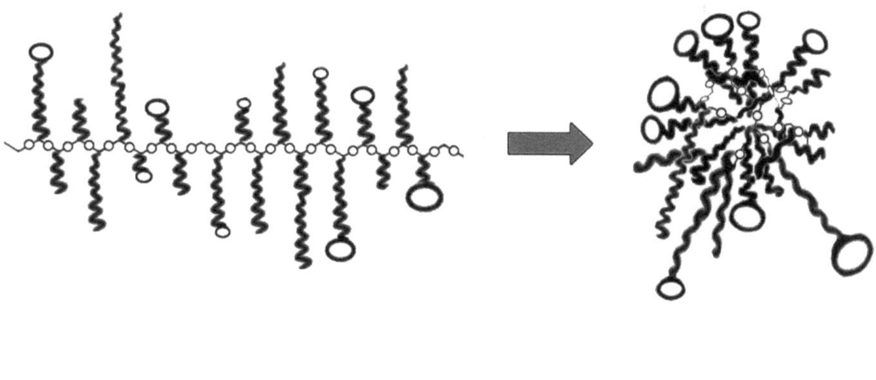

The diagram illustrates the basic idea behind the formation of not-perfect, deformed spherical micelle constructed by 20 different amino acids linked by peptide bonds which determine the sequence of hydrophobicity distribution. Deformation is the effect of significantly limited number of degrees of freedom in respect to freely moving molecules in former examples.

This work addresses protein folding on the basis of two commonly accepted key assumptions:

1. each protein is synthesized to perform a specific biological function, which requires specificity
2. that the abovementioned specificity — the appropriate amount of information - is encoded in the protein's 3D structure.

Consequently, it is useful to apply information theory in the study of protein conformations.

Protein folding remains a challenging problem despite decades of structural biology research [1]. Somewhat paradoxically, identification and analysis of misfolding phenomena may lend useful clues regarding the "correct" course of the folding process.

The folding mechanism, while generic in scope, should also account for the presence of encoded information which determines the protein's intended biological role (Chapter 3). Thus, our research agenda can be subdivided into two core aspects:

1. a generic mechanism rooted in the protein-solvent relationship during folding process,
2. a mechanism of faithfully encoding the specific nature of the resulting protein which is the source of information reversibly directed to water environment.

Based on the analysis presented in Refs. [2—4] which proposes a two-stage folding model, it can be shown that the quantity of information carried by the amino acid sequence is only sufficient to determine the structure of the so-called early stage intermediate (ES), i.e. secondary conformational preferences. Correctly predicting the final form (referred to as the late stage intermediate; LS) — even when the dihedral angles (Φ, Ψ) between each pair of residues are only computed to an accuracy of 5° — requires nearly three times as much information as is initially available. This observation immediately leads to the following question: where does the missing information come from? In addition to identifying its source, also need to determine how that information ensures the protein's specificity. According to the fuzzy oil drop model (3D Gaussian distribution of hydrophobicity in protein body), information required at the late folding stage is contributed by the aqueous environment, which directs the folding process toward the generation of hydrophobic core.

This brings us to the second issue: encoding of specificity.

A generic mechanism, related to the ubiquitous presence of water and its critical importance for biological processes, treats the solvent as the source of an external force field which directs hydrophobic residues toward the center of the emerging protein body, while promoting exposure of hydrophilic residues on its surface. It can be demonstrated that nearly all domains treated as structural units of active proteins conform to this principle (with varying degrees of accuracy). It is also evident that areas where the actual conformation of the protein diverges from theoretical predictions are responsible for ensuring the protein's specificity [5]. A structure which remains in perfect agreement with the model carries relatively little information; however, based on the notion that unlikely events carry more information than frequent ones, we may characterize the aforementioned deviations as the means of expressing information which the protein requires to fulfill its biological role (Chapter 4).

It can also be intuitively noted that a model which relies solely on minimization of internal energy cannot produce a sufficiently diverse spectrum of conformational variabilities to explain the observed complexity of biological processes understood as balance between areas (parts of protein) of low and high information carried. Part of protein body carrying the specificity (information) must locally disobey the universal optimization tendency. The way to detect the presence and quantitatively assess the amount of information coded in is possible using the reference structure which is the ideal micellar hydrophobicity distribution.

The presence of water conditioning biological activity is, however, grossly underappreciated in mainstream protein folding models. The standard approach is to treat the solvent as a collection of particles which interact with the protein's atoms in a pairwise manner. This, however, does not reflect the solvent's structure and its effect upon the polypeptide chain (explicit water models [6,7]). The implicit model expresses the accordance of protein structure with surrounding water transforming the thermodynamically unfavorable exposure of hydrophobic residues into ΔG scale [8,9].

We propose that the solvent should instead be treated as an active and equal "partner" of the protein during folding process. More specifically, we find it desirable to represent the solvent as a continuous external force field whose physiochemical properties are as yet poorly understood (in particular, our knowledge regarding the non-standard dependence of density on temperature what is the critical to support the biological activity — so far only the structure of water ice has been thoroughly described).

In addition to a correct tertiary conformation, which depends upon the presence of water at the folding stage, biological activity of proteins calls for intermolecular interactions. A mechanism by which the protein transfers information to its environment should therefore be identified. Our analysis also addresses this issue — at least to some extent. The fuzzy oil drop model identifies fragments of the protein molecule which arise as a result of interactions with its environment — the so-called accordant fragments (which may span the entire protein), as well as fragments which do not conform to this rule. Quantitative analysis can be performed by assessing the degree to which the presented structure conforms or deviates from the theoretical model. The calculated degree of discordance may also be viewed as a way of transferring information to the environment by enforcing a nonstandard structural arrangement of water particles in a close contact with protein's surface particularly when the hydrophobic area is exposed (Chapter 1).

The fuzzy oil drop model is a convenient way to express active involvement of the environment in generating protein structures. The model presents the extension of oil drop model introduced by W. Kauzmann [10] which was however the discrete model (hydrophobic core and hydrophilic surface). Our model — applying 3D Gauss function to express the distribution of hydrophobicity in protein body appears to be the continuous model (Chapter 2). The model asserts that an external force field drives the folding process in a way which may be compared to the production of a micelle. Micellar structures arise when water acts upon large numbers of bi-polar molecules. Identifying proteins whose structure conforms to the fuzzy oil drop model (i.e. those, which adopt the shape of spherical micelles) lends credence to the model itself (Chapter 2 and 5).

It can be demonstrated that fragments which deviate from the theoretical distribution of hydrophobicity carry more information than accordant fragments. This encoding principle may be interpreted in the scope of tasks which the biological "tool" (i.e. the protein) is expected to fulfill (Chapter 6). Examples of highly disordered proteins, in which the quantity of encoded information is high, will be presented in Chapters 7 and 8.

A separate group of proteins comprises structures which, in order to remain stable, require the permanent presence of specific "chaperones" (Chapter 9) [11]. Such proteins can only attain their intended conformation by contacting another protein, membrane or ligand.

Amyloids represent an extreme outlier – in some respects they should not even be regarded as proteins (Chapter 10). Chemical composition - amino acids – classify them as proteins. Proteins as they appear in nature represent certain well defined biological role in the system called organism. Amyloids fulfill no useful biological role and do not conform to rules imposed by the organism treated as a complete system (for example, they resist degradation). For these reasons amyloids are inherently "alien" and do not belong in biological constructs.

The linear propagation of bands with different level of hydrophobicity which are identified in amyloids are also present in biologically active proteins. The "stop" signal are however observed in these proteins to prevent the potentially unlimited linear propagation (Chapter 11). These examples can be used as patterns for drugs stopping the unlimited propagation observed in amyloids.

The model for proteome construction based on the automatic control system – negative feed back loops - is presented in Ref. [12]. The presence of each compounds is under well defined control system in this model. Amyloids do not belong to such system.

Current work proposes a way of identifying amyloids by searching for linear propagation of alternating bands of high and low hydrophobicity, stretched along the axis of the fibril in contrast to globular proteins with centric concentration of hydrophobicity. Against this background we suggest several possible scenarios of amyloid transformation of transthyretin and immunoglobulin G (VL domain) [13] and transthyretin (Chapter 12). Our conclusions should be treated as a hypothesis which calls for experimental validation. The authors are bioinformaticians and do not have access to laboratory facilities which would permit such validation. We hope that – perhaps –readers who find the presented theories compelling may take it upon themselves to verify their veracity. Clearly, the formation of amyloids must be driven by some kind of force which causes alternating bands of hydrophobicity to arrange themselves into an highly ordered structure. In our view, this driving force is produced by linear clustering of short fragments characterized by similar hydrophobicity, ultimately preventing a shared hydrophobic core from forming. It appears that this effect is not sufficiently counteracted by the aqueous solvent and outweighs the influence of the external force field in its natural form.

As suggested in an earlier edition of this work [14], the ability of some proteins to fold in a highly specific manner, with finely tuned local deviations from the Gaussian distribution enabling them to accommodate a specific ligand or substrate, may be explained by the presence of the ligand during folding. The ligand itself may be viewed as a component of the system: it migrates to its intended location in the 3D Gaussian "capsule", while the protein folds around it, forming a suitable binding pocket. This hypothesis also calls for experimental validation.

In this work we apply the fuzzy oil drop model in the analysis of selected proteins which span the structural spectrum — from near-perfect globules all the way to amyloids. Our goal is to relate the structural properties of amyloids to those of globular proteins.

A globular protein, which contains a (possibly somewhat deformed) hydrophobic core, may be viewed as an intelligent micelle — the "intelligent" part alludes to targeted deviations in the protein's distribution of hydrophobicity, which encode crucial information. In contrast, amyloids contain no information and therefore lack specificity. This point will be illustrated by performing a comparative analysis of the aforementioned spectrum of structural forms — from globules, through disordered proteins (exhibiting varying degrees od discordance) all the way to amyloid fibrils (Chapter 4).

The fuzzy oil drop model describes the protein in terms of a set of parameters. Not all of these are required to fully characterize each structure. However, in order to enable readers to follow the stepwise encroachment of disorder upon protein conformations, we will list all FOD (fuzzy oil drop model) values in each analyzed case.

The presented hypothesis may be criticized on the grounds of our protein selection criteria. The suspicion that proteins have been arbitrarily selected to match our purported confirmation bias may seem convincing, especially in this day and age, where big data analyses predominate. Nevertheless, even individual cases (and note that each case is representative of a whole category of proteins) may support the proposed theoretical model. In the near future we intend to publish tools which we use to identify localized discordances and assess their scope. This will enable readers to apply the presented model to any structures of their choosing. We hope that this will also facilitate further evolution of the model itself.

Diagrams given on the top of each chapter visualize the part described in particular chapter. The best example for 3D Gaussian hydrophobicity distribution is observed in bi-polar surfactant spherical micelle. The parts carrying charge exposed on the surface generate the proper contact with surrounded water environment. Simultaneously the hydrophobic parts of molecules are hidden in the central zone of micelle eliminating disadvantage contact with water. The deterministic character of the orientation of individual molecule in the micelle (due to high symmetry) makes this construction of low-information category.

The structure of co-micelle build by molecules of different bi-polar forms is also easily predictable due to deterministic and highly symmetric construction. The information carried by such system is also low.

Amino acids are also bi-polar with highly differentiated hydrophobicity/hydrophilicity relation. Prediction of the structure constructed by amino acids is significantly more difficult especially taking into account the limitation due to the presence of peptide bonds eliminating free movement of molecules. This is why only limited examples of proteins are able to generate the almost ideal spherical micelle (antifreeze proteins) following the 3D Gauss distribution of hydrophobicity.

Discordance of observed hydrophobicity distribution (in respect to 3D Gauss function) may take very different form. The level of discordance measured quantitatively (Kullback-Leibler entropy) is the way to express the specificity of protein molecule including its biological activity. This is the subject of the Chapter 4.

References

[1] Dill KA, MacCallum JL. The protein-folding problem, 50 years on. Science 2012;338(6110):1042–6. https://doi.org/10.1126/science.1219021.
[2] Jurkowski W, Brylinski M, Konieczny L, Wiśniowski Z, Roterman I. Conformational subspace in simulation of early-stage protein folding. Proteins: Structure, Function, and Bioinformatics 2004;55(1):115–27. https://doi.org/10.1002/prot.20002.
[3] Roterman I, Konieczny L, Banach M, Jurkowski W. Intermediates in the protein folding process: a computational model. International Journal of Molecular Sciences 2011;12(8):4850–60. https://doi.org/10.3390/ijms11084850.
[4] Roterman I, Konieczny L, Jurkowski W, Prymula K, Banach M. Two-intermediate model to characterize the structure of fast-folding proteins. Journal of Theoretical Biology 2011;283(1):60–70. https://doi.org/10.1016/j.jtbi.2011.05.027.
[5] Kalinowska B, Banach M, Wiśniowski Z, Konieczny L, Roterman I. Is the hydrophobic core a universal structural element in proteins? Journal of Molecular Modeling 2017;23(7). https://doi.org/10.1007/s00894-017-3367-z.
[6] Palazzesi F, Salvalaglio M, Barducci A, Parrinello M. Communication: role of explicit water models in the helix folding/unfolding processes. The Journal of Chemical Physics 2016;145(12):121101. https://doi.org/10.1063/1.4963340.
[7] König G, Pickard FC, Huang J, Simmonett AC, Tofoleanu F, Lee J, Brooks BR. Calculating distribution coefficients based on multi-scale free energy simulations: an evaluation of MM and QM/MM explicit solvent simulations of water-cyclohexane transfer in the SAMPL5 challenge. Journal of Computer-Aided Molecular Design 2016;30(11):989–1006. https://doi.org/10.1007/s10822-016-9936-x.
[8] Song B, Charest N, Alexander Morriss-Andrews H, Molinero V, Shea J-E. Systematic derivation of implicit solvent models for the study of polymer collapse. Journal of Computational Chemistry 2017;38(16):1353–61. https://doi.org/10.1002/jcc.24754.

[9] Ható Z, Valiskó M, Kristóf T, Gillespie D, Boda D. Multiscale modeling of a rectifying bipolar nanopore: explicit-water versus implicit-water simulations. Physical Chemistry Chemical Physics 2017;19(27):17816–26. https://doi.org/10.1039/c7cp01819c.
[10] Kauzmann W. Some factors in the interpretation of protein denaturation. Advances in Protein Chemistry 1959;14:1–63. https://doi.org/10.1016/s0065-3233(08)60608-7.
[11] Ellis RJ. Molecular chaperones: assisting assembly in addition to folding. Trends in Biochemical Sciences 2006;31(7):395–401. https://doi.org/10.1016/j.tibs.2006.05.001.
[12] Konieczny L, Roterman-Konieczna I, Spólnik P. The structure and function of living organisms. Systematic Biology 2013:1–32. https://doi.org/10.1007/978-3-319-01336-7_1.
[13] Banach M, Kalinowska B, Konieczny L, Roterman I. Possible mechanism of amyloidogenesis of V domains. Self-Assembled Molecules – New Kind of Protein Ligands 2017:77–100. https://doi.org/10.1007/978-3-319-65639-7_5.
[14] Konieczny L, Roterman-Konieczna I. Conclusion. In: Protein folding in silico; 2012. p. 191–202. https://doi.org/10.1533/9781908818256.191.

CHAPTER 1

Description of the fuzzy oil drop model

Leszek Konieczny[1], Irena Roterman[2]
[1]Chair of Medical Biochemistry, Jagiellonian University — Medical College, Krakow, Poland
[2]Department of Bioinformatics and Telemedicine, Jagiellonian University — Medical College, Krakow, Poland

Contents

This diagram demonstrates the core concept of the fuzzy oil drop (FOD) model, which posits the existence of an external force field generated by the aqueous solvent. The external field guides hydrophobic residues toward the center of the protein body, while hydrophilic residues are instead exposed on its surface. The process continues as the encapsulating ellipsoid capsule shrinks (via changes in σ coefficients of the 3D Gaussian). The gradual increase of gray color visualize the increase of hydrophobicity concentration.

The model has already been presented in detail in numerous publications; nevertheless, we will reintroduce it here in order to provide a theoretical background for further discussions.

The original "oil drop" model, devised many years ago [1] compares the polypeptide to an "oil drop" in the sense that its hydrophobic residues are isolated from the aqueous environment by migrating toward the center of

From Globular Proteins to Amyloids
ISBN: 978-0-08-102981-7
https://doi.org/10.1016/B978-0-08-102981-7.00001-4

the molecule. In parallel, a polar "sheath" emerges to ensure entropically advantageous contact with water. This description may be regarded as a classical discrete qualitative model. As long as hydrophobic residues are entirely isolated from the environment, it appears to work well. If, however, a hydrophilic residue is found to reside at a central location in the protein (or hydrophobic residue exposed on the surface), the model fails to yield accurate results.

In order to mitigate these drawbacks, we have extended the original model, formulating what we refer to as the "fuzzy oil drop" model. This is achieved by introducing a continuous gradient of hydrophobicity, from its maximum value (at the center of the molecule) to near 0 (on the surface). The gradient is mathematically modeled by a 3D Gaussian, which represents the theoretical (or idealized) distribution of hydrophobicity in a perfect protein molecule (i.e. a perfect spherical micelle)—with a prominent central hydrophobic core overlaid by a hydrophilic sheath. Exposure of hydrophilic residues on the surface enables interactions with the aqueous solvent, ensuring solubility. Below we find an intermediate zone, with hydrophobicity increasing along with distance from the surface, depending on the overall volume of the protein body. In a spherical micelle, this gradient is the same in any direction (isotropy), whereas in an elongated globule it may depend on the selected coordinate system axis. The shorter the distance between the surface and the center, the steeper the gradient. The 3D Gaussian is described by three distinct σ coefficients, one for each orthogonal direction. Greater differences between each pair of these values correspond to more elongated globular forms [2,3].

Hydrophobicity is a property of entire amino acid, i.e. a collection of atoms. Thus, for each residue we calculate the position of its so-called effective atom (averaged-out positions of all atoms which comprise that residue). The Gaussian yields a specific value for the location of the effective atom, and this is assumed to represent the theoretical hydrophobicity ascribed to the given residue (T).

The following equation may be used to calculate theoretical hydrophobicity at any point within the ellipsoid capsule:

$$H_i^T = \frac{1}{H_{sum}^T} \exp\left(\frac{-(x_i-\overline{x})^2}{2\sigma_x^2}\right) \exp\left(\frac{-(y_i-\overline{y})^2}{2\sigma_y^2}\right) \exp\left(\frac{-(z_i-\overline{z})^2}{2\sigma_z^2}\right)$$

The point $(\overline{x}, \overline{y}, \overline{z})$is the position of the geometric center of the protein in the 3D coordinate system, when placed in its origin at (0,0,0), these values become 0. The protein should be rotated, making the line linking longest distance between two effective atoms in the molecule coaxial with (say) X-axis. It is then rotated around the X-axis to make the line linking the two most distant positions of the projections of effective atoms on the (say) YZ plane coaxial with Y-axis. Three parameters σ_x, σ_y, σ_z represent standard deviations of the size of the protein, equal to 1/3 of the highest absolute values of x-coordinate, y-coordinate and z-coordinate respectively (according to the 3-sigma rule). The normalizing coefficient H_{sum}^T represents the sum of all H_i^T values of amino acids of the protein, making the H_i^T value normalized. The only input information for the theoretical distribution is a geometrical term concerning the full protein, i.e. the size of the ellipsoid "drop," containing the protein, and characterized by σ_x, σ_y, $\sigma_{z.}$ Traditionally, value of the Gauss function is interpreted as a theoretical idealized hydrophobicity density at given point.

Of course, a real protein is not expected to conform to this model with perfect accuracy. Thus, we calculate the actual (observed) hydrophobicity for each amino acid, which depends on its own intrinsic hydrophobicity (according to any generally accepted scale) as well as on interactions with its neighbors. Following [4], we calculate the hydrophobic interaction assuming a cutoff distance of 9 Å for hydrophobic interactions.

$$H_i^o = \frac{1}{H_{sum}^o}\sum_j \begin{cases} \left(H_i^r + H_j^r\right)\left(1 - \frac{1}{2}\left(7\left(\frac{r_{ij}}{c}\right)^2 - 9\left(\frac{r_{ij}}{c}\right)^4 + 5\left(\frac{r_{ij}}{c}\right)^6 - \left(\frac{r_{ij}}{c}\right)^8\right)\right), & for r_{ij} \leq c \\ 0, \quad for r_{ij} > c \end{cases},$$

Where H_i^o denotes the experimentally observed (index O) hydrophobic density in particular point (position of effective atom of *i-th* residue) which collects the hydrophobic interaction in distance dependent form as given in the formula with the cutoff distance (*c*) assumed according to original work 9 Å [4]. The denominator H_{sum}^o (sum of all H_i^o) makes the value in normalized form. H_i^r and H_j^r express the intrinsic hydrophobicity of *i-th* an *j-th* residues, which can be taken according to arbitrarily selected scale [5,6]. The scale presented in Ref. [6] was taken for calculation in the work discussed here.

Interactions with neighbors may either increase or lower the effective hydrophobicity of each residue—this is reflected by its corresponding observed hydrophobicity value.

Comparison of T and O (following normalization) enables us to unambiguously determine the accordance/discordance between both distributions, either for the protein as a whole, or for selected fragments. We may even point to specific residues as either accordant or discordant versus the theoretical model (O_i vs. T_i).

Given both distributions, it is possible to perform a quantitative comparison by applying Kullback-Leibler's [7] divergence entropy formula:

$$D_{KL}(P \parallel Q) = \sum_i P(i) \log_2 \frac{P(i)}{Q(i)}$$

Where $P(i)$ denotes the observed probability (hydrophobicity density) localized on i-th residue—in this paper called O_i—observed and $Q(i)$ denotes the expected (target distribution) hydrophobicity localized on the same residue—in this paper called T_i—theoretical one—corresponds to the distance between O and T, the latter of which is regarded as the reference.

D_{KL} expresses the formal "distance" between both distributions (T and O). However, since it is constitutes a measure of entropy, the value of D_{KL} cannot be interpreted on its own—a second reference model must be provided. Since T simulates a "perfect" centric hydrophobic core, we may add a reference distribution which lacks any concentration of hydrophobicity at any point in the protein body. This type of distribution—called the unified distribution (R)—assigns hydrophobicity of 1/N to each residue (N being the number of residues in the chain). It represents the status deprived of any form of hydrophobicity differentiation in protein body.

When considering O and R, the value of D_{KL} tells us to what degree the observed distribution approximates the unified distribution. Comparing both values (for O/T and O/R) provides a description of the protein's status: when $O/T < O/R$, the observed distribution is aligned with the theoretical distribution, and therefore the protein may be assumed to contain a hydrophobic core. In the opposite case—$O/T > O/R$—the protein lacks a prominent core.

By applying the 3D Gaussian model and calculating divergence entropy, we obtain a fine-grained description of the protein's status—a procedure which would not be possible under the original "oil drop" model.

In order to avoid having to deal with two distinct values of D_{KL}, we compute another parameter referred to as relative distance (RD):

$$RD = O|T/(O|T + O|R)$$

Where

O|T denotes according to D_{KL} definition

$$O|T = \sum_{i=1}^{N} O_i \log_2(O_i/T_i)$$

And

$$O|R = \sum_{i=1}^{N} O_i \log_2(O_i/R_i)$$

RD (T-O-R) expresses the relation between O and two other distributions—T and R—treated as edge cases. Unlike D_{KL}, this value is independent of the length of the chain and may be used to characterize any protein.

In our search for the causes of discordance between T and O, we have also introduced another type of reference distribution which expresses the intrinsic hydrophobicity of each residue in the input chain. This distribution is labeled H and may be swapped in for R to determine whether the observed structure is dominated by the intrinsic properties of its component residues—as expressed by the corresponding value of RD (T-O-H):

$$RD = O|T/(O|T + O|H)$$

Where O|H according to D_{KL} definition:

$$O|H = \sum_{i=1}^{N} O_i \log_2(O_i/H_i)$$

High values (above 0.5) of RD suggest that no hydrophobic core is present, and additionally that the observed conformation is driven by the "selfish" properties of each residue rather than by the synergistic tendency to produce a shared core. This is why we distinguish RD for (T-O-R) and RD for (T-O-H) relations.

The presented assumptions, when applied to specific proteins, allow us to validate the model as a whole. A full description of the fuzzy oil drop model also needs to acknowledge the correlations between each pair of distributions (T, O and H), expressed as three distinct correlation coefficients

(TvO, TvH and OvH). These coefficients are based on values obtained for selected fragments or for short overlapping fragments of the chain (5 residues each, produced by a moving frame algorithm) and are particularly useful in identifying discordant fragments. Comparing all three coefficients points to specific locations where discordances occur, and also explains their character (by highlighting the causative factor).

In summary, it should be noted that the fuzzy oil drop model is a reflection of a synergistically generated hydrophobic core, which depends on cooperation between residues belonging to the polypeptide chain. Where such cooperation does not occur, the folding process is driven by the intrinsic properties of each residue, which may be regarded as "selfish" action. Under these conditions, the protein cannot reach a globular conformation and alternative structural patterns emerge—including, in some cases, amyloid forms. The specific goal of this study is to suggest a certain path and a mechanism explaining the changes which accompany amyloidogenesis.

Fig. 1.1 provides a graphical depiction of the presented model, along with its interpretation.

Fig. 1.1A visualizes the theoretical distribution, modeled by the 3D Gaussian (T—blue), observed one (O-red) and uniform (R—green). The status of the O distribution in the RD scale is described as discordant versus T distribution. The RD (T-O-R) value equal to 0.693 suggests the closeness versus the uniform distribution. This is why the O distribution is interpreted as lacking the uni-centric hydrophobic core.

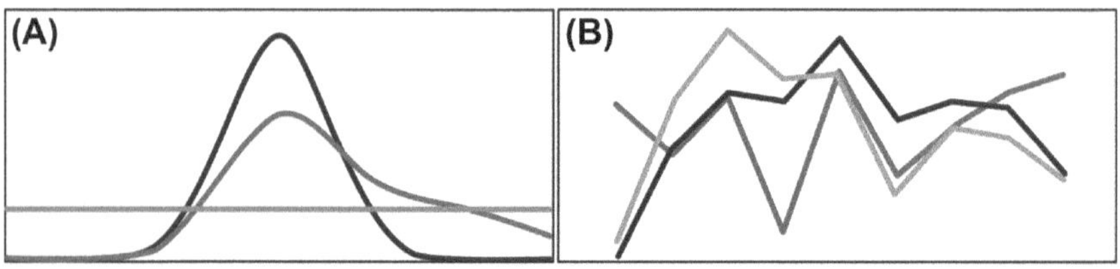

Fig. 1.1 Graphical representation of fuzzy oil drop model parameters reduced to a single dimension for simplicity. (A) theorized Gaussian distribution (T—blue (black in print version)), observed distribution (O—red (gray in print version)) and uniform distribution (R—green (light gray in print version)). (B) theoretical (idealized) hydrophobicity distribution—blue (black in print version), observed—red (gray in print version) and intrinsic hydrophobicity according to the sequence—light blue(light gray in print version). This calculation describes the status of certain polypeptide chain fragment limited to 9 residues.

Fig. 1.1B shows the theoretical distribution (T—blue) for a selected fragment of the chain, while the corresponding observed distribution (O—red) and the intrinsic hydrophobicity values for each residue belonging to the analyzed fragment is shown as H—light blue. The status of discussed chain fragment is expressed by RD (T-O-H) value equal to 0.586. It suggests that the O distribution does not follow the T distribution being dependent on intrinsic hydrophobicity of residues present in the polypeptide chain fragment. The characteristics of selected fragment defines its status in respect to the structural unit (chain or domain).

For the sake clarity, our presentation is limited to a single dimension; however all computations are conducted for the three-dimensional structure (described by the previously mentioned 3D Gaussian).

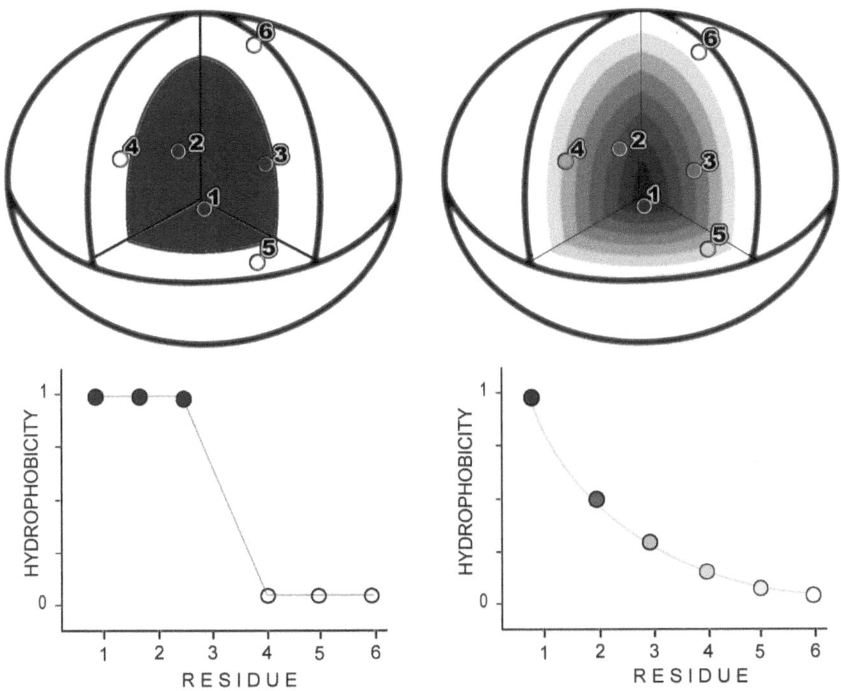

Fig. 1.2 Distribution of hydrophobicity in a protein molecule. Instead of black-white discrete status the continuous one is introduce in fuzzy oil drop model. The positions of residues harmonize their intrinsic hydrophobicity with the expected level of hydrophobicity in protein body. Left -—discrete oil drop model—two layers: black—hydrophobic core; white—hydrophilic shell. Right—fuzzy oil drop model—distribution of hydrophobicity asserted by the model.The bottom profiles express the status of each residue.

The observations described in Refs. [8–11] are treated as supporting the definition of the presented model.

Fuzzy oil drop versus oil drop model

To summarize the description of fuzzy oil drop model the graphic presentation (which appeared in few papers) is recalled. It visualizes the progress from "oil drop"—which is of discrete form to "fuzzy oil drop" of continuous character.

The applicability of fuzzy oil drop model is seen well when the residues do not obey the expected hydrophobicity distribution. Instead of two discrete forms the continuous interpretation is possible allowing measurement of the discordance status (Fig. 1.3).

The interpretation of profiles expressed by fuzzy oil drop model allows distinguishing of local accordance: residues 1, 5, 6, 7 and 9 seem to represent

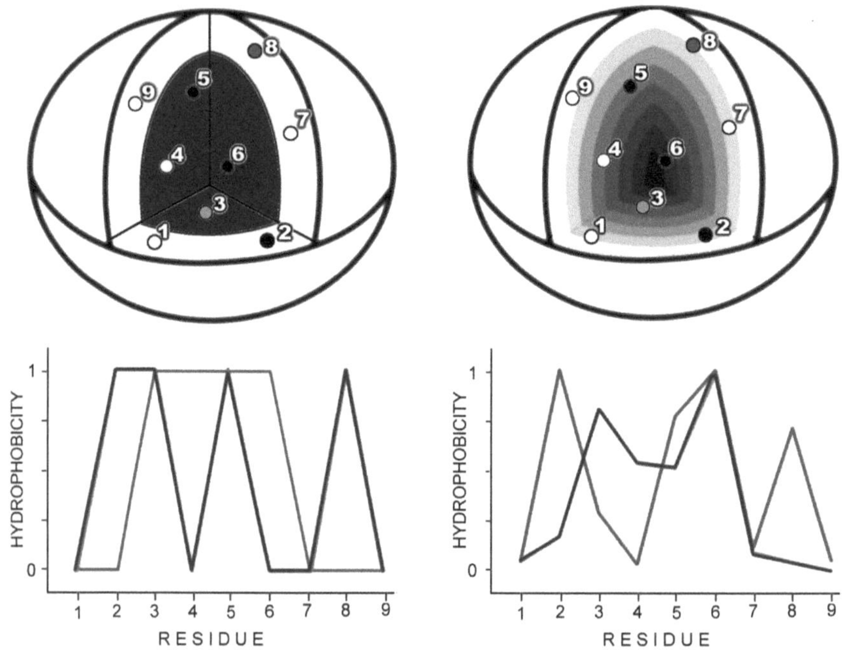

Fig. 1.3 Discordant distribution of hydrophobicity. The positions of some residues cause collision between their status in protein body and the expected level of hydrophobicity. Left—discrete model with distribution of residues not corresponding to the two-layer structure together with profile visualizing the distribution. Right—distribution of hydrophobicity in the continuous model: theoretical (T—blue (gray in print version)) and observed (O—red (dark gray in print version)) together with profile visualizing the distribution. T—blue line (gray in print version), O—red line (gray in print version).

the status accordant with expectation. The residues 3, 4 and 8 can be identified as introducing local discordance.

Amyloids as seen from the perspective of the FOD model

Applying the fuzzy oil drop model to various classes of proteins reveals numerous instances where observed structures closely correspond to theoretical predictions. This is particularly true for individual domains, which in majority contain monocentric cores [8]. Local discordances are frequently associated with biological function—excess hydrophobicity on the protein surface indicates potential complexation sites [9,10], while local hydrophobicity deficiencies often indicate the presence of ligand (or substrate) binding cavities [11].

A particularly interesting case of discordance is observed in amyloids [12]. The observed distribution of hydrophobicity in such structures diverges greatly from the corresponding theoretical distribution, in favor of linear propagation of alternating bands of high and low hydrophobicity parallel to the fibril's axis. Such linear propagation is, by definition, unlimited, enabling unrestricted elongation of the amyloid fibril. In contrast, a prominent hydrophobic core surrounded by a hydrophilic "shell" produces a soluble, globular protein. Fig. 1.4 depicts the structural differences between a globular molecule and a linearly propagating fibril.

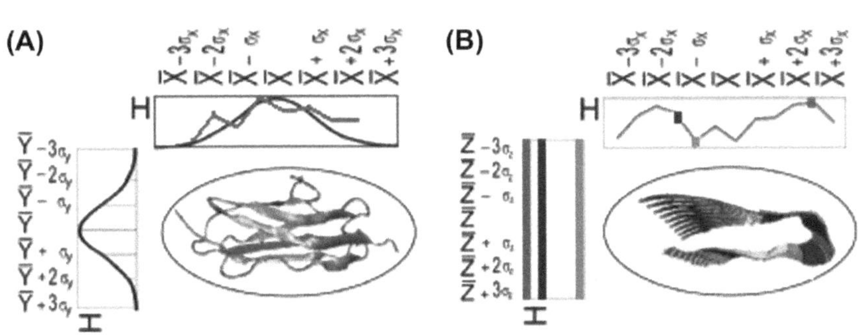

Fig. 1.4 Comparison of a monocentric globular molecule (A) where the observed distribution (red line (gray in print version) representing the status of the fragment distinguished as red in 3D presentation) closely corresponds to the theoretical model (black), and an amyloid (B), which exhibits linear propagation (distribution along the Z-axis perpendicular vs. the plane of the picture of alternating bands of high and low hydrophobicity) [12]. See Ref. [12] for a more detailed introduction to the model. The colors in 3D presentation in B visualize the linear propagation of low/high hydrophobicity (color of points respectively).

It furthermore appears that the linear (not uniform) distribution of hydrophobicity observed in amyloids is closely aligned to the intrinsic distribution (H). One shall distinguish the uniform distribution (called as R) which is isotropic—the constant hydrophobicity is expected in any point of protein body. While the linear distribution expresses the formation of linear bands of different hydrophobicity along and parallel to long axis of the molecule.

In this context, amyloid transformation can be interpreted as optimization of hydrophobic interactions in line with the intrinsic properties of each residue. Minimizing the influence of the aqueous environment may favor such transformation—in contrast to the folding of globular proteins, which depends strongly on active interaction with the aqueous environment (resulting in internalization of hydrophobic residues and exposure of hydrophilic residues on the protein surface) [13-16].

The structural diversity of proteins will be illustrated on the example of titin (good correspondence between T and O) and amyloid Aβ4 (classic linear propagation of alternating bands). Transthyretin appears to represent the middle ground between these two boundary cases. The possible mechanism of transthyretin amyloid transformation is discussed in details in Chapter 12.

References

[1] Kauzmann W. Some factors in the interpretation of protein denaturation. Advances in Protein Chemistry 1959;14:1—63. https://doi.org/10.1016/s0065-3233(08)60608-7.

[2] Konieczny L, Brylinski M, Roterman I. Gauss-function-Based model of hydrophobicity density in proteins. In Silico Biology 2006;6(1—2):15—22.

[3] Roterman-Konieczna I. Protein folding in silico. 2012. https://doi.org/10.1533/9781908818256.

[4] Levitt M. A simplified representation of protein conformations for rapid simulation of protein folding. Journal of Molecular Biology 1976;104(1):59—107. https://doi.org/10.1016/0022-2836(76)90004-8.

[5] Kyte J, Doolittle RF. A simple method for displaying the hydropathic character of a protein. Journal of Molecular Biology 1982;157(1):105—32. https://doi.org/10.1016/0022-2836(82)90515-0.

[6] Kalinowska B, Banach M, Konieczny L, Roterman I. Application of divergence entropy to characterize the structure of the hydrophobic core in DNA interacting proteins. Entropy 2015;17(3):1477—507. https://doi.org/10.3390/e17031477.

[7] Kullback S, Leibler RA. On information and sufficiency. The Annals of Mathematical Statistics 1951;22(1):79—86. https://doi.org/10.1214/aoms/1177729694.

[8] Sałapa K, Kalinowska B, Jadczyk T, Roterman I. Measurement of hydrophobicity distribution in proteins - non-redundant protein data bank. Bio-Algorithms and Med-Systems 2012;8(2):327—38. https://doi.org/10.2478/bams-2012-0023.

[9] Banach M, Konieczny L, Roterman I. Use of the "fuzzy oil drop" model to identify the complexation area in protein homodimers. In: Roterman-Konieczna I, editor. Protein folding. Silico. Oxford, Cambridge, Philadelphia, New Delhi: Woodhead Publishing (currently Elsevier); 2012. p. 95—122.

[10] Banach M, Konieczny L, Roterman I. Can the structure of hydrophobic core determine the complexation site? In: Roterman-Konieczna I, editor. Identification of ligand binding site and protein-protein interaction area. Springer; 2013. p. 41−54.
[11] Banach M, Konieczny L, Roterman I. Recognition of ligand binding binding site. In: Roterman-Konieczna I, editor. Protein folding in Silico. Woodhead publishing (currently Elsevier; 2012. p. 79−93.
[12] Roterman I, Banach M, Konieczny L. Application of the fuzzy oil drop model describes amyloid as a ribbonlike micelle. Entropy 2017;19(4):167. https://doi.org/10.3390/e19040167.
[13] Manoharan VN. Pinned down. Nature Materials 2015;14(9):869−70. https://doi.org/10.1038/nmat4400.
[14] Manoharan VN. Colloidal matter: packing, geometry, and entropy. Science 2015; 349(6251). https://doi.org/10.1126/science.1253751. 1253751−1253751.
[15] Macias-Romero C, Nahalka I, Okur HI, Roke S. Optical imaging of surface chemistry and dynamics in confinement. Science 2017;357(6353):784−8. https://doi.org/10.1126/science.aal4346.
[16] Schutzius TM, Jung S, Maitra T, Graeber G, Köhme M, Poulikakos D. Spontaneous droplet trampolining on rigid superhydrophobic surfaces. Nature 2015;527(7576): 82−5. https://doi.org/10.1038/nature15738.

CHAPTER 2

Folding with active participation of water

Dawid Dułak[1], Małgorzata Gadzała[2], Katarzyna Stapor[3], Piotr Fabian[3], Leszek Konieczny[4], Irena Roterman[5]

[1]Department of Biophysics, Faculty of Physics, Astronomy and Applied Computer Science – Jagiellonian University, Kraków, Poland
[2]Cyfronet AGH Academic Computer Centre CYFRONET - University of Science and Technology in Cracow, Kraków, Poland; Currently - Schibsted Tech Polska Sp. z o. o., Kraków, Poland
[3]Department of Theory of Informatics, Institute of Informatics, Silesian University of Technology, Gliwice, Poland
[4]Chair of Medical Biochemistry, Jagiellonian University – Medical College, Krakow, Poland
[5]Department of Bioinformatics and Telemedicine, Jagiellonian University – Medical College, Krakow, Poland

Contents

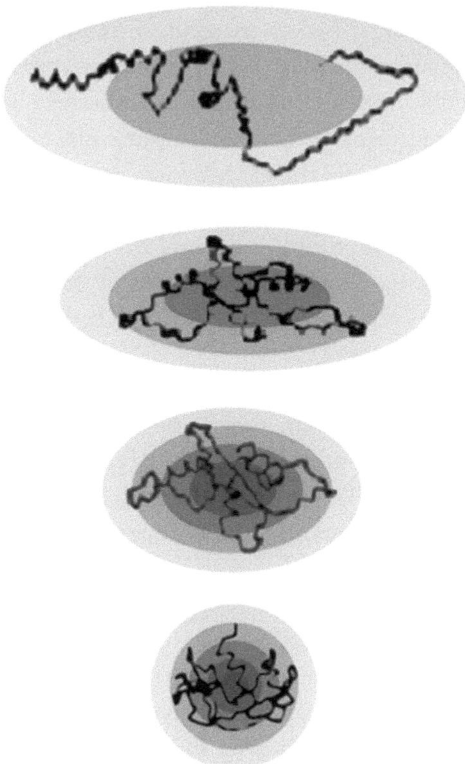

From Globular Proteins to Amyloids
ISBN: 978-0-08-102981-7
https://doi.org/10.1016/B978-0-08-102981-7.00002-6

The diagram illustrates the basic idea behind protein folding simulations based on the fuzzy oil drop model. The Early Stage (ES) intermediate (top picture) of the polypeptide chain is immersed in an external force field represented by a 3D Gaussian. Optimization (minimization) of nonbonding interactions proceeds in parallel with stepwise alignment between T and O, for each residue separately. Minimizing differences between both distributions directs hydrophobic residues toward the center of the emerging structure, while hydrophilic residues are exposed on its surface. The encapsulating ellipsoid slowly shrinks as the folding process progresses. The resulting increase in packing is reflected by a greater gradient of hydrophobicity between the protein's surface and its center.

Several sample proteins have been subjected to folding simulations based on the proposed model. For the sake of simplicity, we selected proteins whose native conformations are consistent with the 3D Gaussian form — i.e. spherical proteins which well-defined hydrophobic cores [1].

Traditional folding algorithms involve minimization of nonbonding interatomic interactions within the protein along with optimization of the corresponding torsion potentials. Water is typically modeled as a pool of external molecules (mono-, bi- or triatomic models) which interact with amino acids in a pairwise manner. Under these assumptions, producing a conformation which exposes hydrophilic residues is often a time-consuming process, requiring intensive computations.

In the fuzzy oil drop model the external force field is simulated by introducing an additional optimization step which reconciles the placement of each residue with the eponymous "fuzzy oil drop" (mathematically represented by a 3D Gaussian) [2]. This step is interleaved with optimization of the molecule's internal energy and can be implemented using the GROMACS software package [3—7].

The Gaussian is constructed individually for each iteration to match the existing intermediate structure. In successive iterations the capsule shrinks until an acceptable degree of packing has been attained (depending on nonbonding interactions). The final result is then evaluated using CASP similarity metrics, enabling quantitative comparisons between models and their corresponding targets [8].

ES (Early Stage) — the staring structure is generated according to ES model described in details in Refs. [9,10].

Fig. 2.1 visualizes successive steps of internal free energy optimization (labeled E) followed by optimization of hydrophobic forces (labeled H). The corresponding stepwise changes in FOD status are shown in Fig. 2.2,

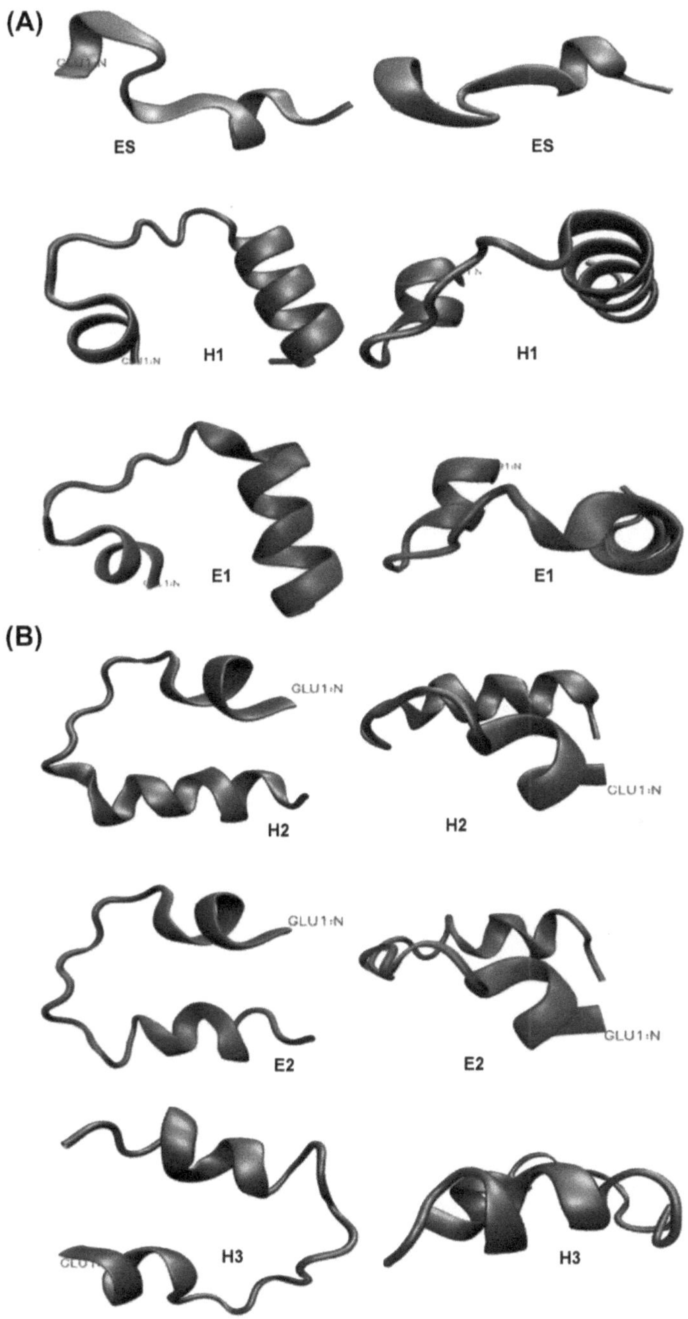

Fig. 2.1 Sample de novo protein (1FME) in successive iterations of the folding process. For each iteration the corresponding internal energy value is listed in Fig. 2.2. (as calculated by GROMACS [3]). Symbols E and H correspond to internal energy and hydrophobicity optimization steps respectively.

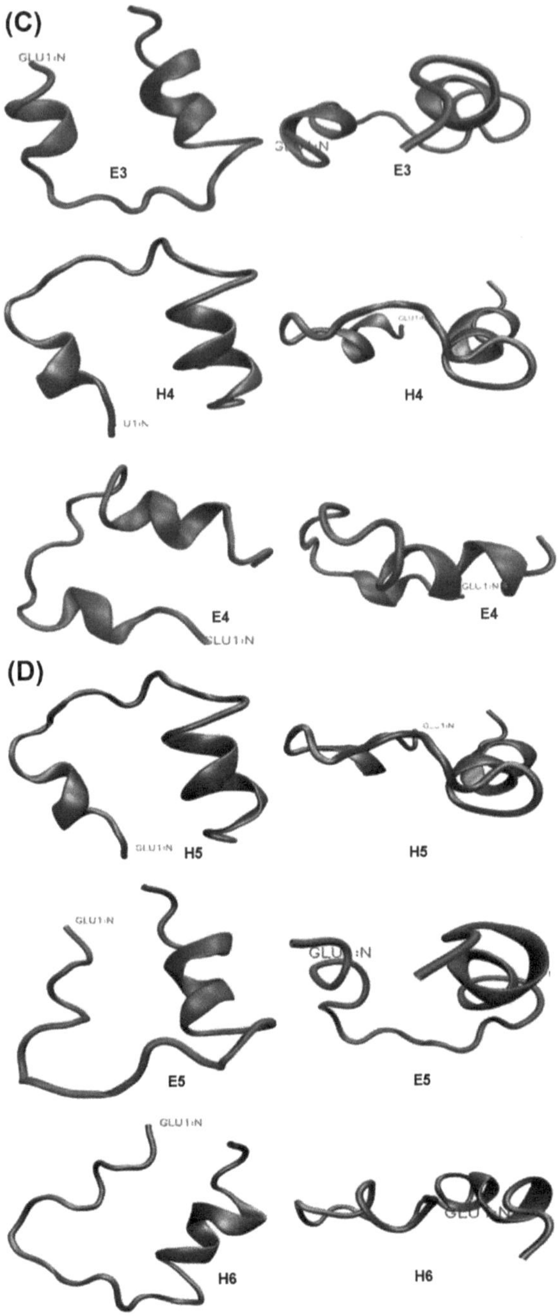

Fig. 2.1 (*continued*).

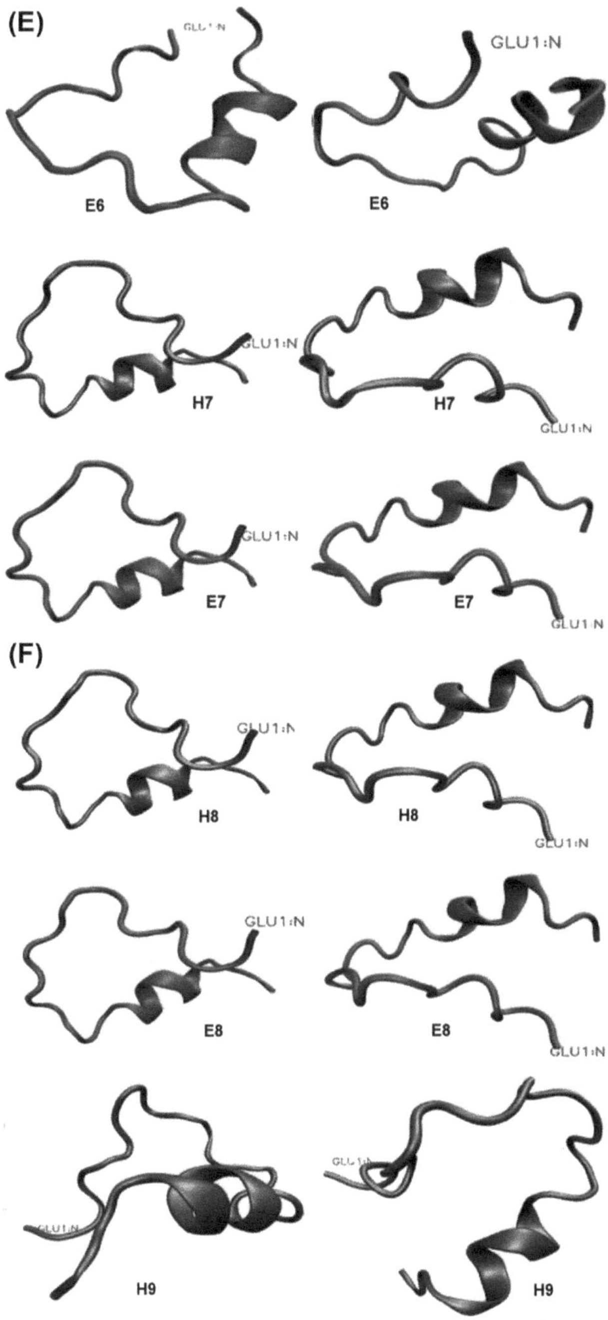

Fig. 2.1 (*continued*).

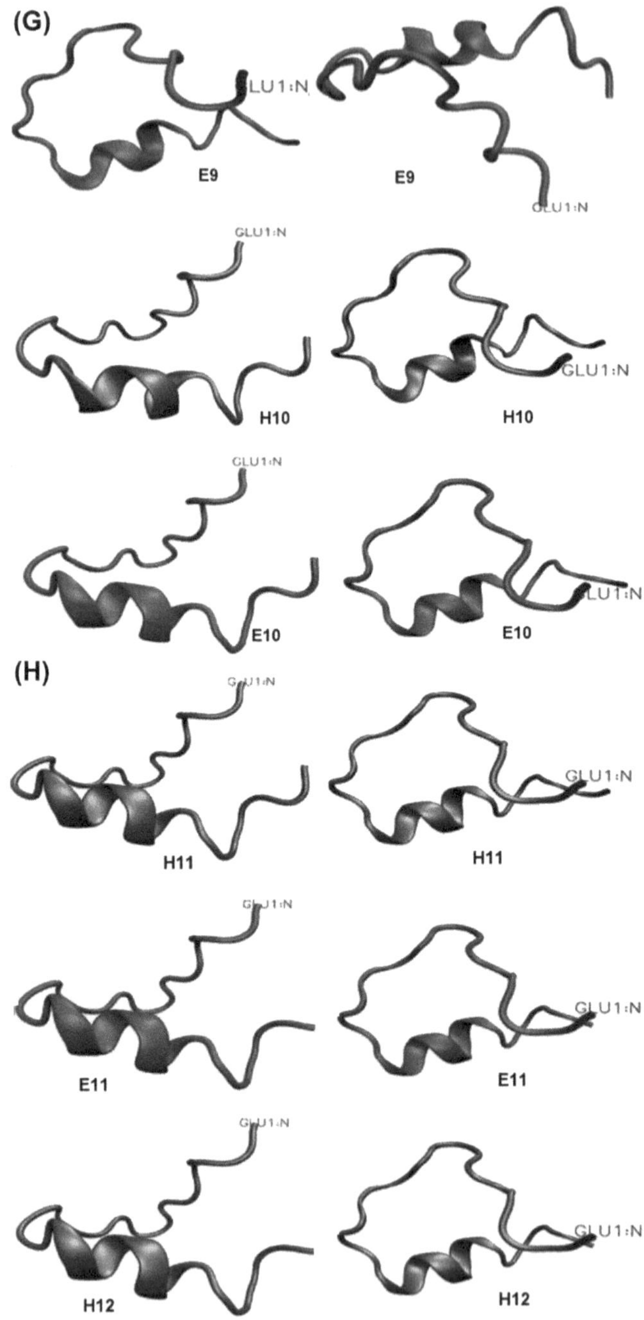

Fig. 2.1 (*continued*).

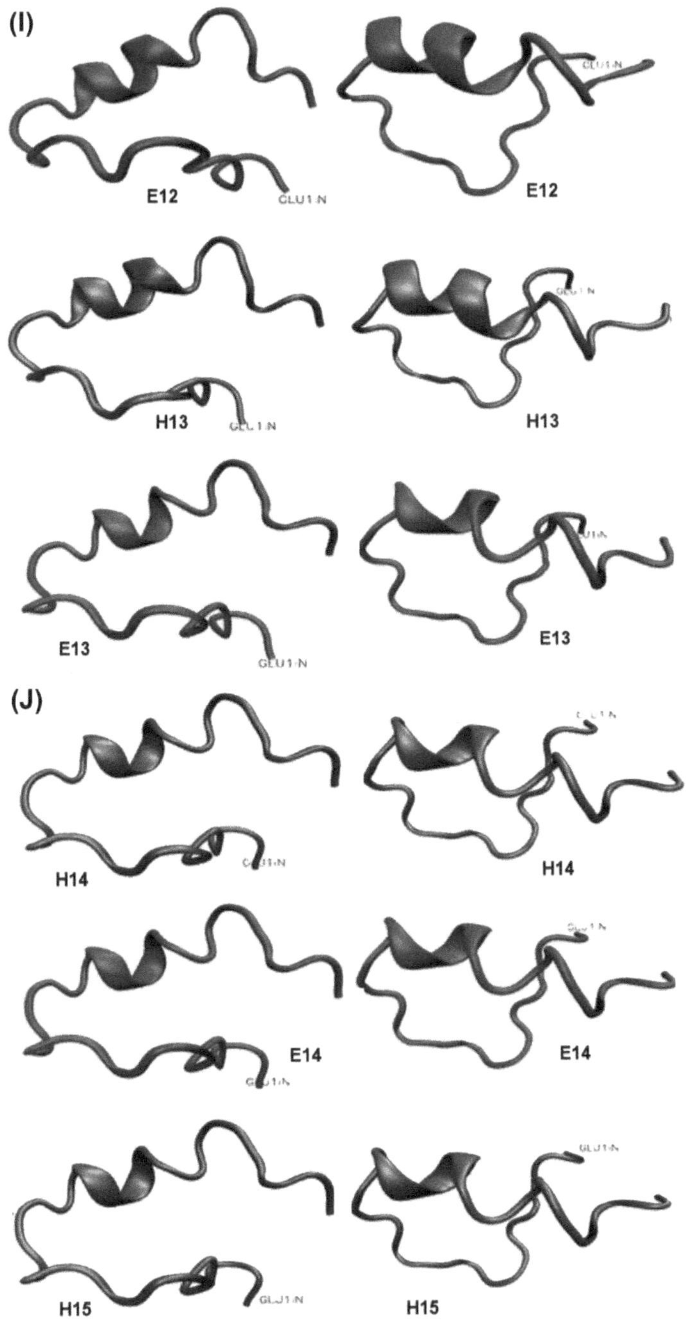

Fig. 2.1 (*continued*).

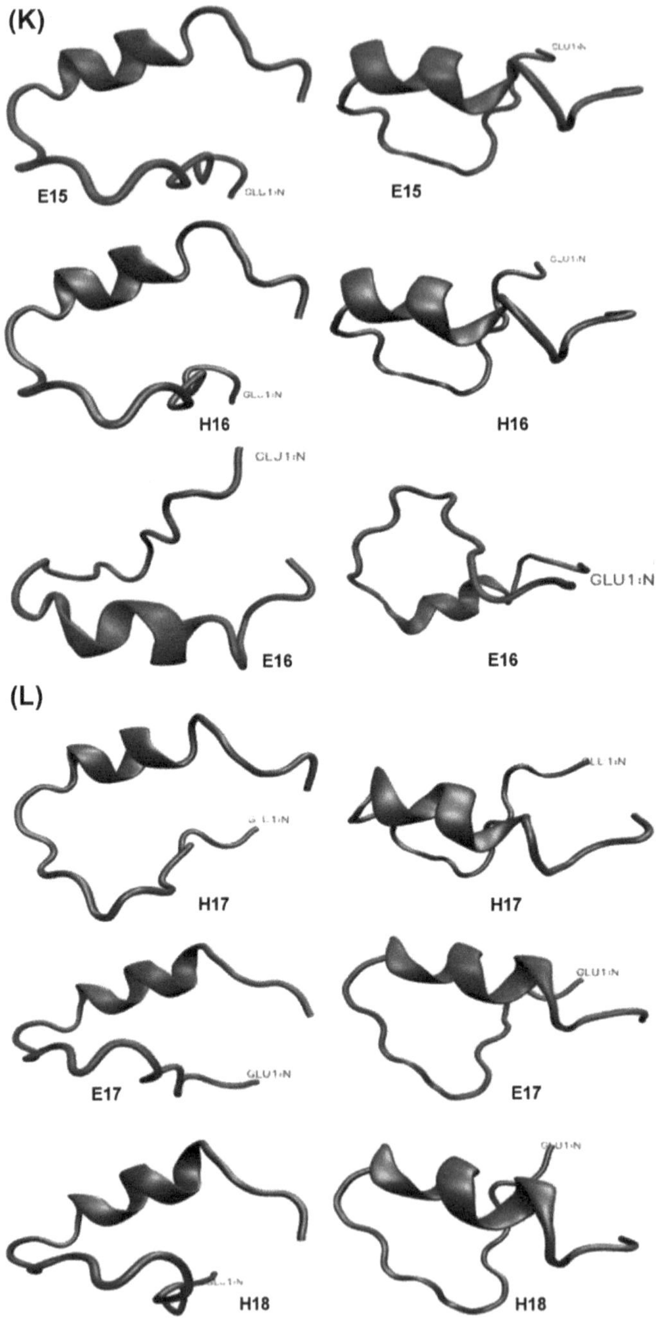

Fig. 2.1 (*continued*).

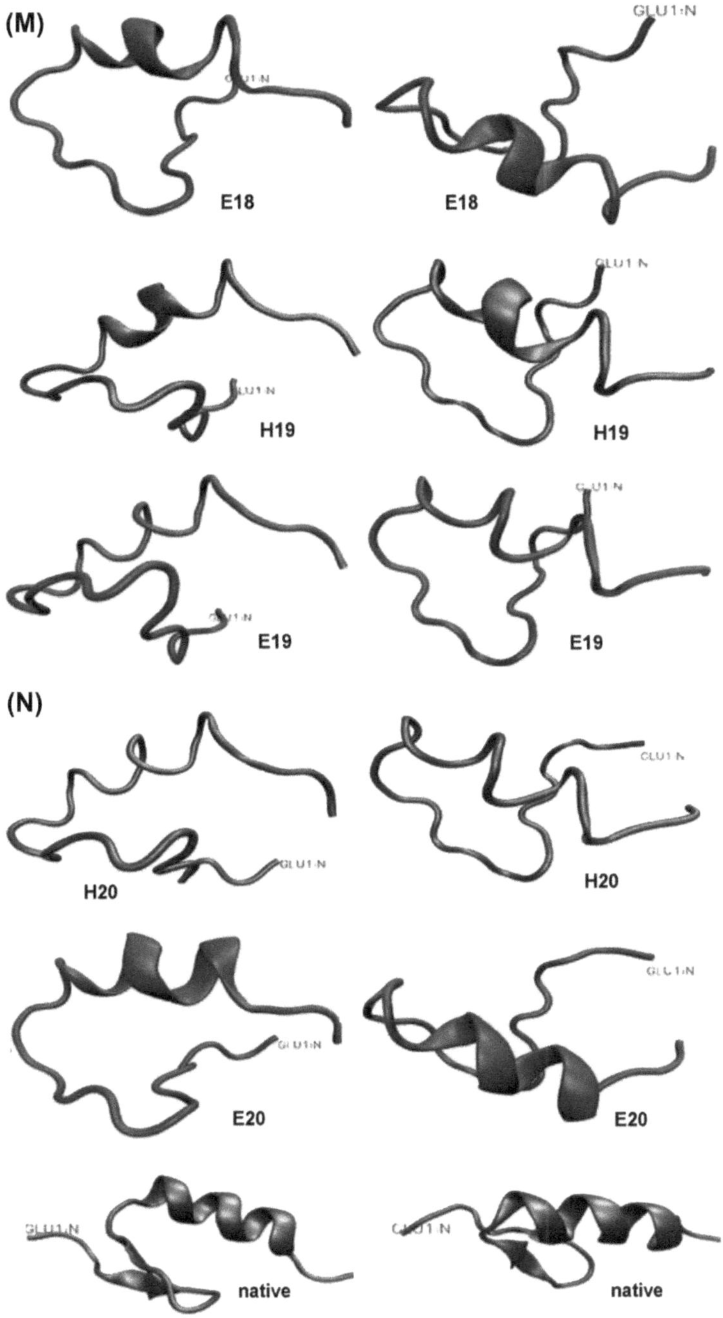

Fig. 2.1 (*continued*).

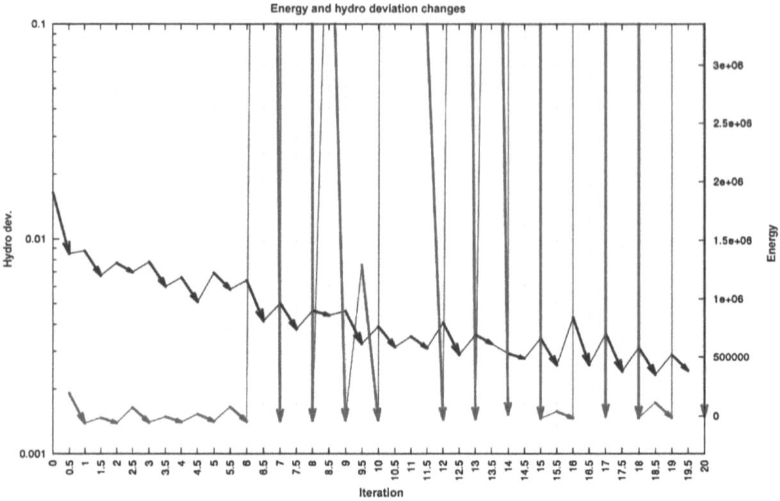

Fig. 2.2 Changes accompanying successive iterations of the folding process. Red line: internal free energy optimization (according to GROMACS); blue line: fuzzy oil drop model fitting (axis called Hydrophobicity). The chart corresponds to structures shown in Fig. 2.1.

while Fig. 2.3 presents the alignment between the resulting model and the target.

The presented example shows that a hydrophobic core emerges as the protein adopts a micelle-like conformation. Other examples of protein folding simulation are also presented in Chapter 10.

The profiles illustrated in Figs. 2.2 and 2.3 reveal a contradiction between optimization of free energy and optimization of hydrophobic interactions. In light of this fact, it seems useful and reasonable to end the process with a final internal energy optimization step. It is due to the fact, that the hydrophobicity optimization is performed on effective atoms. Thus it may introduce the inter-atomic collisions when the structure is transformed to all atom model. Fig. 2.2 reveals progressive formation of a hydrophobic core. It seems that the best alignment is obtained in the 18th iteration, where RMS-D reaches its lowest value, while GDT is almost at its maximum level Fig. 2.3.

The final structures shown in Fig. 2.4 differ somewhat from the target, particularly with regard to their N- and C-terminal fragment. Fig. 2.5 reveals a concentration of hydrophobic residues at the center of the protein, along with exposure of hydrophilic residues on its surface. The target structure, 1FME, is characterized by the following parameters: RD(T-O-R):

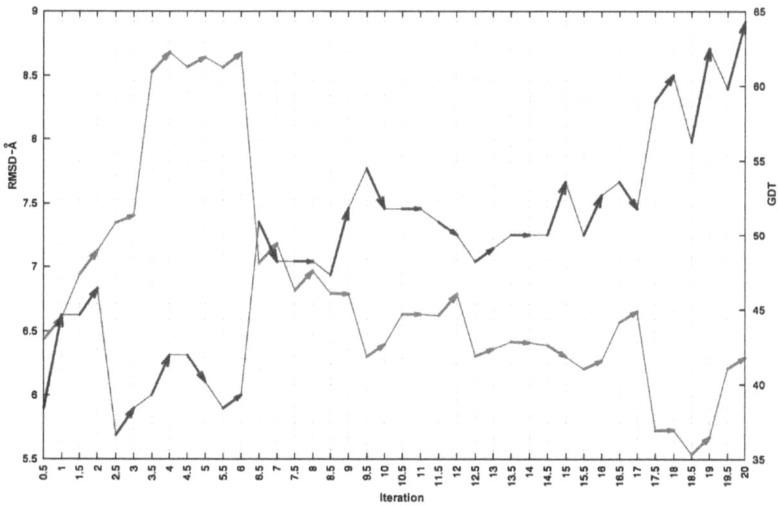

Fig. 2.3 Changes in RMS-D (Å) and GDT (no unit – parameter used in CASP project to assess the similarity between target and model [8]) during folding. The chart corresponds to structures shown in Fig. 2.1.

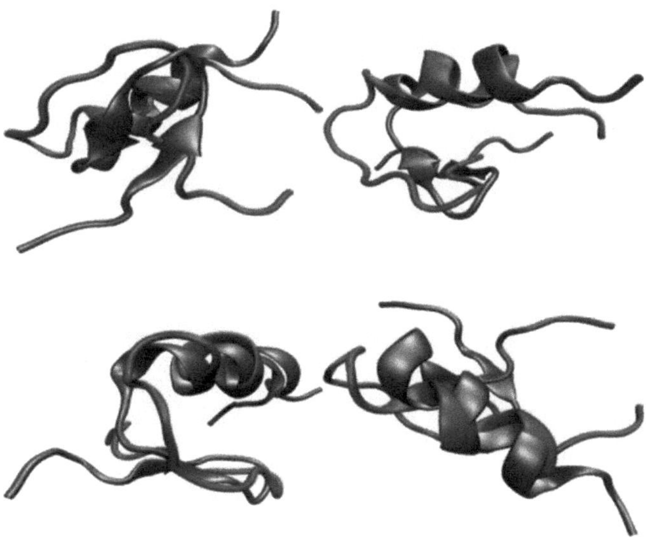

Fig. 2.4 Final models (pink) superimposed onto the native structure of de novo protein (1FME).

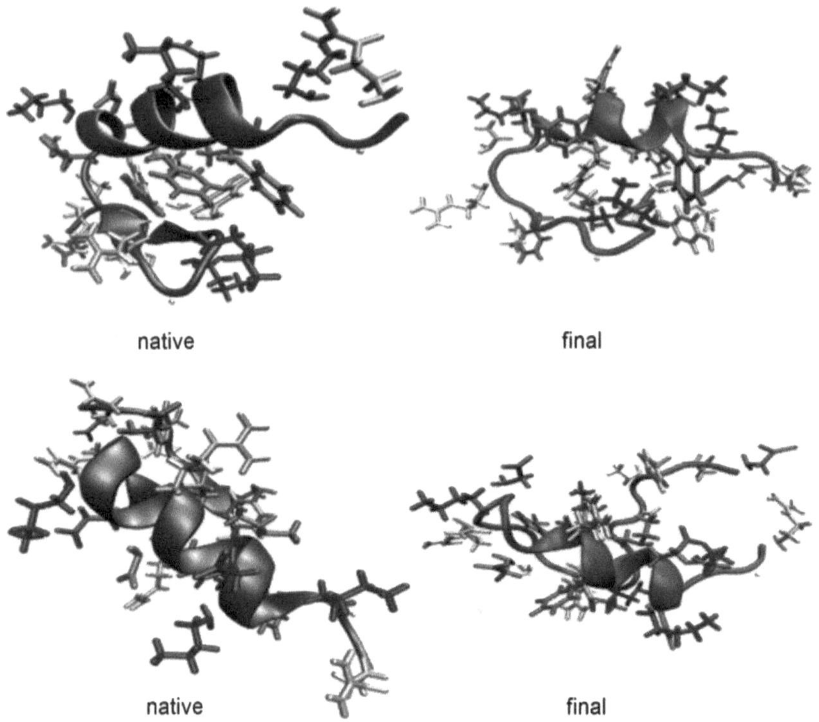

Fig. 2.5 Native and final structures showing hydrophobic (red) and hydrophilic (blue) residues (1FME).

0.298; RD(T-O-H): 0.249; HvT: 0.706; TvO: 0.821; HvO: 0.863. Consequently, this protein is regarded as highly accordant with the theoretical model and represents a useful test case. Our model, generated by the folding simulation, has RD values of 0.294 and 0.089 (T-O-R and T-O-H respectively), along with the following correlation coefficients: HvT: 0.460; TvO: 0.770; HvO: 0.609.

Another sample folding simulation based on the presented model is discussed in Ref. [2].

The authors are currently developing a software package where interleaved optimization of internal (nonbonding interactions) and external (fuzzy oil drop model) force fields is replaced by a multicriteria optimization procedure representing a tradeoff between both factors. Introduction of the Pareto front approach is expected to more accurately reflect the balance between internal and external force fields.

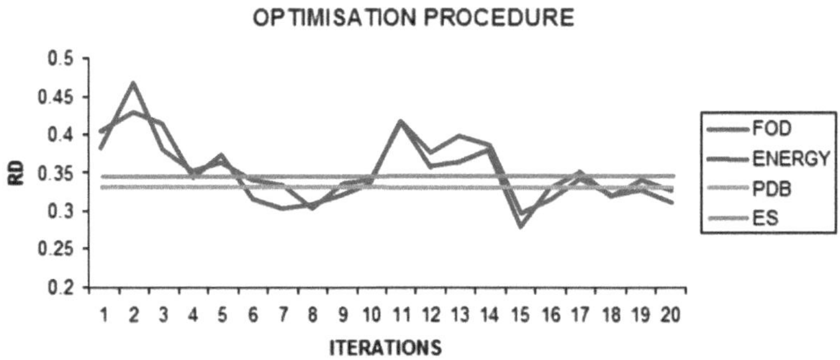

Fig. 2.6 Changes in RD (the measure of idealized – 3D Gauss function distribution) and observed hydrophobicity distribution) during folding. The chart corresponds to structures shown in Fig. 2.1. FOD – status after FOD optimization, Energy – status after energy optimization, PDB – status as it appears in crystal structure, ES – status of early stage model.

To follow the role of FOD influence on folding process the changes of RD are shown in Fig. 2.6.

The energy values represents the internal energy status after the Gromax iteration. The values described as FOD represent the status of the structure after FOD iteration. The convergence can be seen especially in the final steps of procedure. The line described ES represent the status of Early Stage structure generated according to Refs. [9,10]. The line identified as PDB visualize the status of the structure available in PDB. There are 34 models available. The range of RD for all models is as follows: 0.306–0.356. The line represents the averaged RD value for all 34 models treated as reference form.

It would also be highly desirable to validate the model with a broader set of proteins and simulation parameters (e.g. number of iterations).

References

[1] Konieczny L, Brylinski M, Roterman I. Gauss-function-Based model of hydrophobicity density in proteins. In Silico Biology 2006;6(1–2):15–22.

[2] Gadzała M, Dułak D, Kalinowska B, Baster Z, Bryliński M, Konieczny L, et al. The aqueous environment as an active participant in the protein folding process. Journal of Molecular Graphics and Modelling 2019;87:227–39. https://doi.org/10.1016/j.jmgm.2018.12.008.

[3] van Gunsteren WF, Berendsen HJC. Groningen molecular simulation (GROMOS) library manual. 1987. p. 1–221.

[4] van Gunsteren WF, Berendsen HJC. Computer simulation of molecular dynamics: methodology, applications, and perspectives in chemistry. Angewandte Chemie International Edition in English 1990;29(9):992–1023. https://doi.org/10.1002/anie.199009921.

[5] van Gunsteren WF, Billeter SR, Eising AA, Hunenberger PH, Kruger P, Mark AE, et al. Biomolecular simulation: the GROMOS96 manual and user guide. Zurich: Verlag der Fachvereine& Berendsen, H. J. C; 1996.
[6] http://www.gromos.net.
[7] van Gunsteren WF, Bakowies D, Baron R, Chandrasekhar I, Christen M, Daura X, et al. Biomolecular modeling: goals, problems, perspectives. Angewandte Chemie International Edition 2006;45(25):4064–92. https://doi.org/10.1002/anie.200502655.
[8] http://predictioncenter.org.
[9] Jurkowski W, Brylinski M, Konieczny L, Wiśniowski Z, Roterman I. Conformational subspace in simulation of early-stage protein folding. Proteins: Structure, Function, and Bioinformatics 2004;55(1):115–27. https://doi.org/10.1002/prot.20002.
[10] Jurkowski W, Baster Z, Dułak D, Roterman-Konieczna I. The early-stage intermediate. Protein Folding in Silico 2012:1–20. https://doi.org/10.1533/9781908818256.1.

CHAPTER 3

Information encoded in protein structure

Leszek Konieczny[1], Irena Roterman[2]
[1]Chair of Medical Biochemistry, Jagiellonian University – Medical College, Krakow, Poland
[2]Department of Bioinformatics and Telemedicine, Jagiellonian University – Medical College, Krakow, Poland

Contents

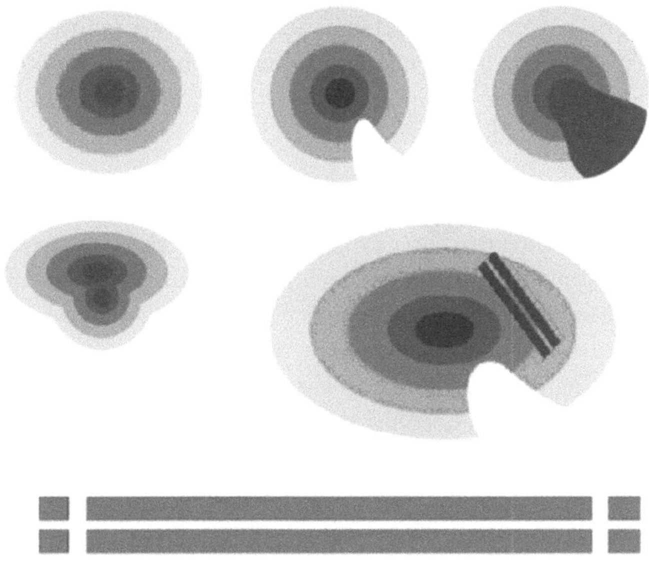

Schematic depiction of variable quantities of information encoded as specific deformations in a protein "micelle" (of varying complexity), producing many different conformations – from spherical to ribbon-like. Except for the first and last structure, each form in the sequence encodes information in proportion to its deviation from

From Globular Proteins to Amyloids
ISBN: 978-0-08-102981-7
https://doi.org/10.1016/B978-0-08-102981-7.00003-8

the theoretical distribution of hydrophobicity. Such deviations may be regarded as a way to ensure the protein's specificity, (note, however that the first structure — i.e. a spherical micelle — and the last structure — a ribbonlike micelle — are devoid of information and therefore nonspecific).

The prevailing dogma which assumes that the protein's conformation is fully encoded in its sequence, seems to underestimate the role of external factors which affect protein folding.

A basic definition, due to Shannon, specifies the quantity of information (bit) carried by an event with probability p_i as:

$$I_i = -\log_2 p_i$$

This formula may be applied to information processed by biological systems. Given four types of nucleotides which encode for 20 amino acids, it is clear that — in order to unambiguously identify a specific amino acid — at least three nucleotides are required. A simple calculation then reveals that a surplus of information exists on the side of nucleotide triplets. The transfer of information between nucleotides and amino acid sequences is therefore relatively straightforward. The same, however, is not true for the transition between amino acid chains and 3D structures. The 3D structure itself may be treated as a specific means of encoding information which is crucial for the protein to perform its function. A protein's 3D structure can be accurately described in terms of dihedral angles (Φ and Ψ) between each pair of adjacent residues.

As it turns out, the quantity of information required to unambiguously define each angle with an accuracy of at least 5 degrees, is two to three times greater than the quantity carried by each amino acid. More specifically, the information content of a single residue is on the order of 4–6 bits (depending on its frequency of occurrence). In turn, the amount of information necessary to define the pair of Φ and Ψ angles with a precision of 5 degrees is 8–11.5 bits (taking into account the conformational preferences of each residue).

As shown above, there is a notable deficiency of information on the side of the amino acid chain — however, if we restrict our search to identifying secondary conformational characteristics, the available information might prove sufficient. Thus, instead of the full Ramachandran plot, we confine our search to a specific subspace, represented by an elliptical path which traverses all areas corresponding to well-known secondary folds [1,2]. This process reduces our demand for information and indicates that the input chain may indeed provide enough information to determine a conformation — but only of the early-stage intermediate (ES). Consequently, we divide the folding process into several intermediate stages, as follows:

$$U \rightarrow ES \rightarrow LS \rightarrow \text{native 3D}$$

U — unfolded structure (plain amino acid sequence); ES — early-stage intermediate restricted to the limited conformational subspace (details in Refs. [2,3]); LS — late-stage intermediate leading to the native (3D) form of the protein.

Introducing two intermediate steps decreases the amount of information required at each stage. Restricting our search to a limited conformational sub-space, i.e. an elliptical path on Ramachandran map (see Refs. [2,3] for a detailed description), facilitates identification of **Φ** and **Ψ** angles (with a 5-degree step along the elliptical path) for the ES intermediate. It appears that the amount of information required to define these values of **Φ** and **Ψ** corresponds to the information content of the amino acid chain. This implies that the quantity of information carried by the amino acid sequence is only sufficient to determine the structure of ES. This intermediate is very important since it embodies the greatest challenge faced by protein structure prediction algorithms. Attempts to define starting structural forms (ES intermediate) in protein structure prediction models follow many diverse approaches, including the following:

1. comparative modeling: querying structure databases for 3D forms whose corresponding sequences are a good match for the given input sequence, and then applying genetic algorithms to further align the structures of both models. This method is often limited to homological proteins and can be described as evolution-based.
2. compiling databases of short structural motifs which may be assembled into a starting structure which is then subjected to further modeling. For example, the Robetta software uses lists of 3- and 9-residue fragments [4],
3. optimizing the conformational space for rapid searching by limiting the degrees of freedom available for each rotation [5],
4. simplifying the input structure by reducing it to a coarse-grained form — much like the preceding method, this operation also reduces the dimensionality of the conformational space [6].

Traversing the full conformational subspace has recently become a feasible option thanks to major advancements in IT and computer science; however this method is sometimes criticized as being out of touch with experimental realities. Clearly, *in vivo* protein folding cannot rely on a brute-force approach, since it would then require far more time than is actually the case.

As previously noted, the information carried by the residue sequence, regardless of the applied theoretical approach, is not sufficient to unambiguously define the native 3D form of the resulting protein. A simple corollary is that progressing from ES to LS calls for an additional source of information. In our view this source is represented by the aqueous environment. It should be noted that the presence of water is an indispensable condition of proper protein folding — a fact so obvious that its full implications are often overlooked.

Attempts to factor the aqueous solvent into *in silico* folding simulations have a long history [7]. A popular approach is to introduce a certain number of water molecules (depending on solvent density) into the bounding box which contains the target chain. In this technique, the polypeptide interacts with the solvent in a pairwise fashion, i.e. through a network of atom-atom interactions (note that water may be modeled in various ways, as mono-, bi- or tri-atomic molecules). The simulation then takes into account nonbonding interactions (electrostatic, van der Waals, torsion potentials and H-bond potential) [4—6,8].

It appears, however, that the abovementioned procedure does not accurately capture the impact of the aqueous environment. Restricting analysis to a set of interactions between pairs of atoms neglects the holistic influence of the solvent upon the resulting 3D structure. Therefore, the fuzzy oil drop model disposes with this schema in favor of a continuous force field, treated as a "background" for interactions which occur between the polypeptide's constituent atoms. The specific nature of this external field depends on the structural properties of water (which, as yet, are poorly understood — at least for the liquid phase); however in all cases it promotes internalization of hydrophobic residues and formation of a hydrophobic core. A classical theory constructed on the basis of this assumption is the so-called "oil drop" model [9], which predicts the existence of a hydrophobic core encapsulated by a hydrophilic shell. In its basic version the model is discrete, i.e. it only recognizes two possible states (hydrophobic center + polar surface); however despite this drawback it accurately captures the role of the environment in terms of isolating hydrophobic residues from contact with water. We may speculate that similar results could be obtained using the pairwise interaction method; however, this would likely entail an arduous and lengthy simulation process.

In the course of our work we have proposed a continuous extension of the discrete oil drop model, which we refer to as the "fuzzy oil drop" model (Chapter 1 and 2). It enables us to quantitatively define the structural ordering of the protein's hydrophobic core, as well as to assess any potential deviations from theoretical predictions. As noted in Ref. [10], many biologically active proteins are highly consistent with the 3D Gaussian distribution of hydrophobicity, lending support for the model itself. Analysis of a nonredundant set of protein structures derived from PDB further indicates that a vast majority of individual domains adhere to the model with high accuracy [10].

Many protein families, including the so-called downhill and fast-folding proteins [11,12], fold in accordance with the fuzzy oil drop model. This shows that the quantity of information present in the amino acid sequence is sufficient to determine its 3D structure — but only in the presence of an additional source of information, i.e. the solvent. The fast-folding family is particularly noteworthy in this respect: its members are capable of rapidly reverting to their native forms regardless of how many times they have been unfolded. This ability to "automatically" assume the intended tertiary conformation in the absence of any other stimuli clearly shows that missing information comes from the solvent (even without assessing the specific quantity of bits which the solvent imparts to the protein).

For the reasons stated above we feel confident in stating that the ES-to-LS folding stage draws information from the protein's environment, and that therefore the environment plays an active and crucial role in the folding process. Interestingly, experimental studies report that undesirable changes in environmental properties (changes in pH, ionic potential etc.) have a detrimental effect on protein folding — an observation which is fully consistent with our proposed theory.

The equivalence between the quantity of information fed into the folding process (residue sequence; aqueous environment) and the information content of the resulting 3D structure is not so much computed as experimentally demonstrated by reversibility of folding/unfolding. Still, this equivalence does not fully resolve the question of how proteins attain their intended conformations. The presented approach describes a general process without referring to the function of specific proteins. In some cases no function-oriented structural changes are necessary — for example, antifreeze proteins (which are structurally highly consistent with the fuzzy oil drop model) perform their function simply by being dispersed in the solvent where they can disrupt the formation of ice crystals. Such proteins do not need to attract or bind to any external structures since this would restrict their exposure to water.

In most cases, however, the schematic folding process presented above omits a crucial issue: the way in which the protein's 3D structure encodes its intended function:

$$U \rightarrow ES \rightarrow LS \rightarrow 3D \rightarrow FUNCTION$$

This additional stage — from 3D structure to function — may also be analyzed on the basis of information theory concepts.

A hypothetical protein which is fully consistent with the 3D Gaussian distribution of hydrophobicity would exhibit two important properties: perfect solubility and inability to interact with any other molecules (except for surface-bound ions). In actual proteins whose activity calls for interaction with external structures, certain local deviations from the theoretical distribution of hydrophobicity are expected. As shown in the chapter devoted to ligand binding and protein complexation (Chapter 6), the nature of this phenomenon is highly complex. Most proteins are very selective and therefore need to deviate from the theoretical distribution of hydrophobicity in a very specific and controlled manner. Thus, the quantity of information needed to produce a protein with a specified activity profile must be greater than the quantity needed to produce an inert protein (or a protein whose activity is entirely determined by its solubility). This, again, calls for an additional source of information.

In our previous book which introduces the fuzzy oil drop model, we postulated [13] that the enzyme must fold in the presence of its target substrate. The substrate would need to take an active part in the folding process — the enzyme effectively folds "around" the substrate, which automatically generates a suitable binding cavity. Note, however, that this assumption remains speculative and calls for experimental validation.

There are, however, other potential sources of information, such as chaperones: proteins, which work by temporarily attaching themselves to the target polypeptide chain and modulating the external force field which guides the folding process. It is also worth noting that any such modulations are local in scope and produce similarly local distortions in the resulting structures.

The quantity of information carried by a chaperone may be precisely calculated by comparing the structure of the chaperone-assisted protein with the hypothetical conformation which it would reach in the absence of the chaperone. Kullback-Leibler's divergence entropy coefficient, D_{KL}, provides a *de facto* quantitative measure of this difference [14].

Can a protein be accurately referred to as a chemical molecule? Or, in other words — what differentiates a protein from any other compound, whether organic or inorganic? Saying that "the protein is synthesized by a living organism" does not fully address the problem. Instead, we propose a definition which refers to the targeted nature of proteins. The protein's native form is not a goal unto itself, and the folding process does not end when the protein attains its intended 3D structure — rather, the process may be considered complete only after the protein has gained biological function. In this sense, the term "folding" relates not to the protein's structure, but to its biological role. If the protein has not folded in the intended manner, it is degraded and disposed of: from the biological point of view it is useless and cannot be considered a "biological molecule", even though it may have been synthesized by a living organism.

In short, the protein is a tool which must perform a specific task. It would therefore be misleading to assume that the 3D structure is the ultimate goal of the folding process.

The sources of information required to produce specific local deviations from the ideal distribution of hydrophobicity are varied and depend on the complexity of the structure which must be produced to perform the given task. Validating the correctness of simulated structures makes sense only if the protein in question has been proven to perform its function. Consequently, RMS-D scores should not be regarded as the sole criterion of the reliability of structure prediction algorithms [15,16].

Many evolutionarily conditioned processes which produce desirable results may be explained by invoking the following formula, well known to information scientists:

$$P = [1 - (1 - p)k]$$

P — overall probability of successfully completing a task; p — probability of success for a single attempt; k — number of attempts.

There are two ways to increase P (i.e. the odds of successfully completing a given task). One way to approach this problem is to maximize k, i.e. the number of attempts, each of which may succeed with probability p. A classic example is a lottery where each contestant may purchase an arbitrary number of tickets. Purchasing more tickets (greater k) gives one better chances of winning the lottery (greater P). This method is most often applied when the player cannot deduce the correct solution *a prori* (i.e. when the winning numbers are not known to them). The downside is that it entails a significant expenditure of resource and energy to produce the large number of "coins" whatever its form is.

The other approach is to try to increase the value of p. In the case of a numbers game we may achieve this effect by investigating how the winning numbers are drawn, and by introducing a deterministic factor (for example, if the drawing machine uses numbered balls, we may surreptitiously insert small magnets into certain balls and a larger magnet into the drawing tube, thus increasing the likelihood that our designated balls will be selected). In this case, the expenditure of resources is negligible, but the process instead calls for additional information – specifically, information regarding the inner workings of the selection process. In short, we must be aware of how the system works and also know the desired outcome (set of balls), both of which require information.

In biology a classic mechanism which relies on high values of k is plant pollination. Since a plant is unable to determine the optimal placement of seeds to ensure germination, plants instead produce vast quantities of seeds and must bear the associated costs (energy expenditures). Similar solutions are employed by the reproductive systems of animals, including mammals. A sperm cell does not know how to locate the egg – it therefore faces a similar problem to a plant trying to disseminate its offspring.

In contrast, some plants produce rhizomes, which represent an alternative way of increasing the likelihood of successful proliferation – in this case, by increasing the value of p. The rhizome is essentially a living laboratory, which takes care of all of the plant's vital processes. A robust rhizome serves as proof that the plant's requirements are well taken care of. Such rhizomes are more likely to sprout a new plant. When digging out a plot of perennials we may sometimes find wilted, stunted rhizomes which have not encountered suitable conditions and cannot initiate proliferation – or even support themselves. Thus, the condition of the rhizome provides additional information which is required to increase the value of p. In this particular case the required quantity of information is vast and involves all vital processes which take place in the rhizome. Notably, the rhizome also encodes the intended outcome of the selection process: finding a place which promotes growth of its offspring. A human analogy would be *in vitro* fertilization – thanks to in-depth knowledge of the reproduction process and also of its expected outcome, we may create a new organism using just a handful of sperm cells.

On the molecular level, the "increased k" approach is embodied by synthesis of IgG antibodies. This process proceeds without knowledge of the intended target, i.e. the antigen which may have entered the organism. Modifications of CDR fragments increase the likelihood of obtaining a combination which matches a particular (but unknown) antigen. The greater the diversity of IgG antibodies the greater the chance of initiating a successful immune response. Of course, the flipside of this process is that it consumes a great deal of energy, needed to sustain synthesis of a vast array of proteins which differ with respect to their CDR fragments. In contrast, vaccination represents a way to tackle the problem by increasing the value of p, since the antigen (representing the given disease) has already been recognized and may be introduced into the organism in a controlled manner, to develop immunity.

Approaches based on increasing p are also employed in systems which encode information related to the specific nature of their associated processes, such as all enzymatic catalysis reactions.

The quantity of information encoded in any specific system varies along with the system's complexity and its subdivision into stages. For example, enzymatic active sites, responsible for catalysis and admitting the presence of water, are usually found on the molecular surface or in shallow pockets. If, however, a given reaction requires an anhydrous environment, the system becomes far more complex since the active site must accommodate its intended substrate while at the same time excluding water. Examples of highly complex systems which rely on the "increased p" approach are provided by all enzymes (or receptors) which have a quaternary structure. This layer of the structural hierarchy resolves two problems: it facilitates construction of an active site characterized by high information content (complicated structure), and it also provides a way to prevent unintended initiation of a given process. The latter property is ensured by the complex nature of the receptor itself: if any one of its components is missing, the process cannot begin. In some cases, the number of required components (and therefore the quantity of required information) is very high. An extreme example is provided by the ribosome, which, on the one hand, must know how to carry out protein synthesis, while on the other hand must be able to validate that all conditions for the synthesis of a given protein have been met (i.e. that all required components are present). This complicated process also provides a way to exert tight control over protein synthesis, which is fundamentally important for all organisms.

The visualization of the P dependence on p and/or k is shown in Fig. 3.1.

The complex form of a ribosome (as wall as of any receptor which has a quaternary structure) encodes a lot of information. In theory, it might be possible to pack the necessary information into a single polypeptide chain. In practice, however, synthesizing a complex active site becomes far easier

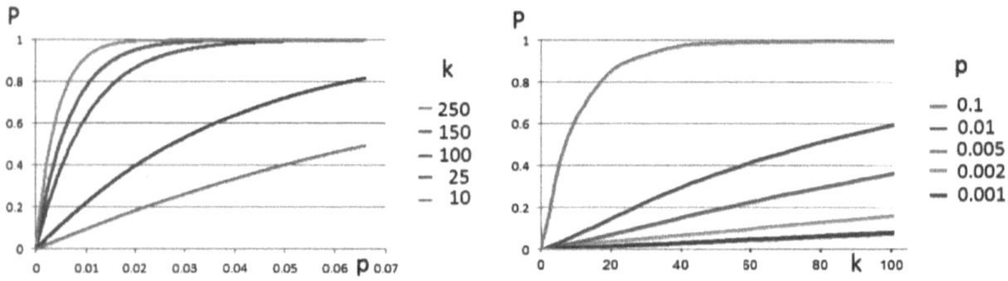

Fig. 3.1 Association between the probability of reaching the goal (*P*) and: (Left) the number of attempts (*k*) for a variable elementary probability (*p*), (Right) the probability associated with each elementary event (*p*) for a variable number of attempts (*k*).

when the given structure is assembled from smaller fragments, each of which contributes some of the necessary information. This assembly provides a way to combine small pieces into a single information-rich system whose operation may be compared (on information theory grounds) to a logical conjunction of events. Notably, conjunction carries more information than a logical alternative (which would be difficult to implement within the bounds of a single biological structure) — this is why the active site is often located in closeness, proximity to the molecular (or domain) interface, or even forms part of that interface. Each structural unit contributes a piece of information required to trigger a complex and highly specific process.

Enzymatic catalysis, in addition to providing the necessary conditions for a given reaction to occur, must overcome one additional obstacle: the need for intermolecular communication. The protein's structure (including its quaternary structure) must, in addition to enabling catalysis, also send out specific signals to attract potential partners, i.e. other molecules involved in the given reaction. Such short-range intermolecular communication needs to make use of the aqueous environment. This issue will be further discussed in Chapter 7, where we investigate the effects exerted by the protein upon the surrounding solvent.

Let us refer once again to the definition of a protein as a tool required to perform a specific and potentially highly complex task. A chemical molecule obtained through artificial synthesis (whether organic or inorganic) lacks an inbuilt purpose — it simply exists. The same is true for a protein devoid of biological activity: it may be regarded as a standalone entity, not associated with any process and not fulfilling any goal. Such anomalous proteins, often resulting from mutations, are undesirable and, in most cases, disposed of by the organism. A particularly interesting example of an aberrant protein is supplied by amyloids (naturally, we refer to pathogenic amyloids, rather than to amyloid-like molecules which the organism can make use of — the differences between both classes are presented in chapter 7).

A properly folded protein, capable of performing its intended task (whatever it may be), encodes information related to that task. In some cases, the protein's function implies interaction with another molecule, membrane or cell — the nature of the interaction partner must therefore also be encoded.

Can a micelle function as a carrier of information? A surfactant micelle, much like a micellar protein (e.g. a type III antifreeze protein), emerges as a result of immersion in an aqueous environment, which determines its structure through selective interaction with polar groups. The micelle therefore exemplifies a response to an external force field, generated by water. It encodes as much information as is necessary to progress from ES to LS. This information is encoded in the structure of the monomer (surfactant) — particularly in its size and polarity gradient, both of which determine its reaction to the external force field.

The structure of a micelle may be characterized as a passive adaptation to external forces, explaining the similar solubility of surfactant micelles and type III antifreeze proteins [17,18].

How, then, should we define an amyloid? Amyloids are essentially anomalous ribbon-like or cylindrical [19] micelles, which emerge not as a *result* of external forces, but *in spite* of them. The aqueous solvent does not control the process which leads to formation of amyloid fibrils — rather, the fibril actively opposes the influence of the solvent. The source of information may be found in the intrinsic properties of constituent residues rather than in their environment. This observation may be regarded as paradoxical, but it explains the peculiar properties of amyloids. In their case, information is carried by the sequence — or, more accurately, by the intrinsic hydrophobicity of its constituent residues. It appears that amyloid structures emerge when the effect of the external force field is diminished, for instance by a change in the physiochemical properties of water. This corresponds to a change in the properties of the external field, enabling intrinsic hydrophobicity to guide the folding process to a conclusion which differs from "natural" conditions. Notably, physiochemical changes in the solvent have been found to promote amyloidogenesis [19].

Shaking is a known "nonchemical" inducer of amyloid formation. In physical terms, shaking alters the properties of the solvent through aeration, which, in turn, increases the interphase boundary area. The folding process is also hindered in the presence of detergents — likely not through direct interactions with the polypeptide chain but rather due to a change in the properties of water [19].

To conclude our study of the role of information in protein folding/misfolding, we may remark that the aqueous environment — when modeled as a continuous force field — provides information crucial for proper folding of polypeptide chains and ensuring that they are capable of fulfilling their biological role. Changes limited to the properties of this field (e.g. presence of urea) may result in protein disability. Very often the process is reversible and in the absence of the denaturing factor the solvent reverts to its natural state, where proper folding may again occur. Unfortunately, however, amyloids do not follow the above rules: once formed, they do not undergo structural modifications when environmental conditions change.

The above hypothesis constitutes a core aspect of the presented study.

References

[1] Alejster P, Jurkowski W, Roterman-Konieczna I. Structural information involved in the interpretation of the stepwise protein folding process. Protein Folding in Silico 2012:39—54. https://doi.org/10.1533/9781908818256.39.

[2] Jurkowski W, Brylinski M, Konieczny L, Wiíniowski Z, Roterman I. Conformational subspace in simulation of early-stage protein folding. Proteins: Structure, Function, and Bioinformatics 2004;55(1):115—27. https://doi.org/10.1002/prot.20002.

[3] Jurkowski W, Baster Z, Dułak D, Roterman-Konieczna I. The early-stage intermediate. Protein Folding in Silico 2012:1—20. https://doi.org/10.1533/9781908818256.1.

[4] Kim DE, Chivian D, Baker D. Protein structure prediction and analysis using the Robetta server. Nucleic Acids Research 2004;32(Web Server):W526—31. https://doi.org/10.1093/nar/gkh468.

[5] Kolinski A, Skolnick J. Monte Carlo simulations of protein folding. I. Lattice model and interaction scheme. Proteins: Structure, Function, and Genetics 1994;18(4):338—52. https://doi.org/10.1002/prot.340180405.

[6] Krupa P, Hałabis A, Żmudzińska W, Ołdziej S, Scheraga HA, Liwo A. Maximum likelihood calibration of the UNRES force field for simulation of protein structure and dynamics. Journal of Chemical Information and Modeling 2017;57(9):2364—77. https://doi.org/10.1021/acs.jcim.7b00254.

[7] Science (n.d.). doi:10.1126/science.

[8] Van Gunsteren WF, Berendsen HJC. Computer simulation of molecular dynamics: methodology, applications, and perspectives in chemistry. Angewandte Chemie International Edition in English 1990;29(9):992—1023. https://doi.org/10.1002/anie.199009921.

[9] Kauzmann W. Some factors in the interpretation of protein denaturation. Advances in Protein Chemistry 1959;14:1—63. https://doi.org/10.1016/s0065-3233(08)60608-7.

[10] Sałapa K, Kalinowska B, Jadczyk T, Roterman I. Measurement of hydrophobicity distribution in proteins - non-redundant protein data bank. Bio-Algorithms and Med-Systems 2012;8. https://doi.org/10.2478/bams-2012-0023.

[11] Roterman I, Konieczny L, Jurkowski W, Prymula K, Banach M. Two-intermediate model to characterize the structure of fast-folding proteins. Journal of Theoretical Biology 2011;283(1):60—70. https://doi.org/10.1016/j.jtbi.2011.05.027.

[12] Kalinowska B, Banach M, Wiśniowski Z, Konieczny L, Roterman I. Is the hydrophobic core a universal structural element in proteins? Journal of Molecular Modeling 2017;23(7). https://doi.org/10.1007/s00894-017-3367-z.
[13] Kullback S, Leibler RA. On information and sufficiency. The Annals of Mathematical Statistics 1951;22(1):79–86. https://doi.org/10.1214/aoms/1177729694.
[14] http://predictioncenter.org.
[15] Gadzała M, Kalinowska B, Banach M, Konieczny L, Roterman I. Determining protein similarity by comparing hydrophobic core structure. Heliyon 2017;3(2):e00235. https://doi.org/10.1016/j.heliyon.2017.e00235.
[16] Roterman I, Banach M, Konieczny L. Antifreeze proteins. Bioinformation 2017; 13(12):400–1. https://doi.org/10.6026/97320630013400.
[17] Banach M, Konieczny L, Roterman I. Why do antifreeze proteins require a solenoid? Biochimie 2018;144:74–84. https://doi.org/10.1016/j.biochi.2017.10.011.
[18] Roterman I, Banach M, Konieczny L. Application of the fuzzy oil drop model describes amyloid as a ribbonlike micelle. Entropy 2017;19(4):167. https://doi.org/10.3390/e19040167.
[19] Serpell LC. Alzheimer's amyloid fibrils: structure and assembly. Biochimica et Biophysica Acta (BBA) - Molecular Basis of Disease 2000;1502(1):16–30. https://doi.org/10.1016/s0925-4439(00)00029-6.

CHAPTER 4

Globular or ribbon-like micelle

Leszek Konieczny[1], Irena Roterman[2]
[1]Chair of Medical Biochemistry, Jagiellonian University—Medical College, Krakow, Poland
[2]Department of Bioinformatics and Telemedicine, Jagiellonian University—Medical College, Krakow, Poland

Contents

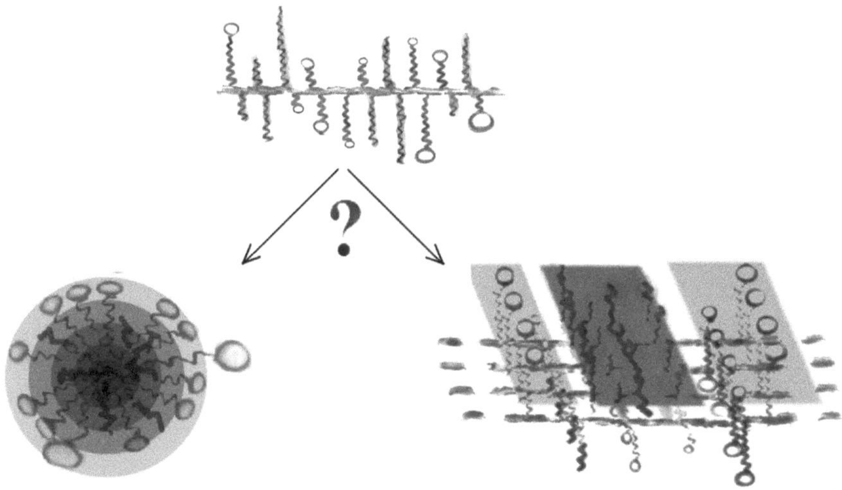

Taking decision: SPHERICAL (globular) or RIBBON-LIKE micelle (fibrilar) ?

As already remarked in the Introduction—in this publication we treat the protein as a specialized tool suited for a particular task. While there are no *useless* proteins in the organism, the role fulfilled by each protein

From Globular Proteins to Amyloids
ISBN: 978-0-08-102981-7
https://doi.org/10.1016/B978-0-08-102981-7.00004-X

may be either simple or complex. For this reason, proteins vary in terms of their information content.

The term "specific" used in biochemistry is equal to saying "information" and in particular "information package" to express the equipment which is able to ensure the appearance of biological activity.

The second assumption used in based on the Heisenberg's uncertainty principle, which excludes equal precision for measurement of two canonically conjugate variables. If the structure of protein is assessed on the level of individual atoms the description of the entire structural unit (for example domain) is impossible—particularly its biological activity is hard to be defined. Function (treated as the aim-oriented local cumulation of information) can be identified using the complete structural unit as a whole.

Particularly hydrophobicity can not be defined using individual atoms as objects for analysis.

It is the characteristics of the set of atoms—like amino acids in particular.

The relation of protein molecule can be expressed solely and exclusively in relation to water. The aim-oriented coding of biological function can be treated in respect to symmetry (order). The higher symmetry the lower information coded. The presence of local discordance in respect to overall order of the object is the form of coding system expression.

Benzene can be taken as an example. As long as all H atoms are linked to C atoms in the aromatic ring any substitution reaction is highly difficult. The consequence of one H atom substituted by for example $-NH_2$ (nitro-benzene) is an easy substitution of the second H atom in ring in a well defined position −meta. The presence of $-NH_2$ is treated as local discordance carrying the information as to the next reactivity position.

Information content is intimately linked with the notion of symmetry. In symmetrical systems, individual components are well ordered and follow a predetermined structural pattern. An example of a (complex) chemical structure characterized by strong symmetry is a spherical micelle. Individual particles with specific chemical properties (polar/nonpolar fragments) self-assemble into a micellar structure when exposed to water. The process is deterministic: given a unit molecule, we may accurately predict the conformation of the resulting micelle. All the required information is encoded in each molecule (note, however, that not all molecules are capable of generating micelles). The driving force, which enables this information to spontaneously manifest itself, is the presence of water.

It is also possible to accurately model the structure of a co-micelle, which comprises more than one type of molecule. The specific nature of each participating molecule carries information, which, based on intermolecular dynamics and interactions with the water environment, determines the structure of the so-called co-micelle. Given sufficient knowledge of the structural properties of each unit molecule, we may model the resulting co-micelle with high accuracy.

Effect of the aqueous environment on intramolecular processes

The polypeptide chain is a combination of 20 different molecules with different predisposition to generate the spherical micelle. Since amino acids differ in terms of their hydrophobicity profiles, the protein may be treated as a peculiar co-micelle. Nevertheless, it is important to point out a major difference between proteins and surfactant micelles: the presence of covalent bonds between each pair of residues which form the "protein micelle." These bonds severely restrict the conformational freedom of the polypeptide chain—unlike in surfactant micelles where the number of degrees of freedom is much greater.

Even so, we may assume that the protein during folding process follows the mechanism of micelle-forming, though one incapable of achieving an ideal distribution of hydrophobicity. The consequence of disability to create the idealized spherical micelle proteins contain local deformations, which are due to two factors: the aforementioned limitations imposed by the presence of covalent bonds, and the differing physiochemical properties of amino acids, which represent the building blocks of micellar proteins.

Let's assume for the moment that it is possible to construct a perfect micelle, with an ideal centrally placed hydrophobic core and a regular hydrophilic outer shell. We may ask—what benefit would such a structure bring to the organism? By itself, the micelle carries no information, other than that it is highly soluble and does not interact with other molecules (both properties are in fact closely related). It might be suspected of having no use whatsoever—but, surprisingly, there is a role for such molecules in living organisms (as discussed in Chapter 5). We refer here to type III antifreeze proteins—near-perfect micelles which perform an important function in organisms capable of surviving subzero temperatures. Their solubility is an important factor, since antifreeze proteins must be evenly distributed in the aqueous environment. Exposure of polar groups on their surface alters the structural properties of water in such a way as to prevent formation of ice crystals. Thus, they remain useful—despite carrying almost no information.

If we were to sort proteins by information content, antifreeze proteins type III would need to be placed immediately after the classic surfactant micelle.

In information theory, information may be conflated with the occurrence of an unlikely event. Similarly to surfactant micelles, the structure of a type III antifreeze protein—which carries very little information—may be accurately predicted. The same is true for the so-called fast-folding (or ultra-fast-folding) proteins, which exhibit very strong accordance between the observed (O) and theoretical (T) distribution of hydrophobicity. In some cases, these turn out to be mutated proteins where mutations confer the ability to fold in a very rapid manner.

Moving on to proteins which carry appreciable quantities of information, we note that the structure of such proteins must include elements whose probability of occurrence is low (in structural terms), even though the remainder of the protein may still follow the theoretical (Gaussian) distribution of hydrophobicity with high accuracy. This is a general rule, which results from interaction with the aqueous solvent. As the environment promotes formation of protein micelles, structural deviations which encode specificity must counteract this tendency.

Significant reduction of degrees of freedom of amino acids in chain in relation to freely moving surfactant molecules and determined neighborhood (determined sequence) eliminates the possibility to generate the spherical micelle. The simple one chain enzyme may be the example. The only discordant area in the protein molecule is the active site cavity.

Eliminating residues which comprise such cavities reveals that the remainder of the molecule still resembles a well-ordered micelle and may be generated in a spontaneous manner. Evolution appears to have selected the 20 biogenic amino acids (characterized by differing hydrophobicity profiles) to enable generation of "imperfect" structures, whose imperfections (when compared to surfactant micelles) encode information.

The specific activity of proteins results from the predetermined balance of order and disorder within their micellar structure, and from the associated local deformations.

Along with enzymes, the group of proteins whose information content may be described as modest includes ligand-binding proteins which carry prosthetic groups. Attracting and binding a ligand requires a deformation which corresponds to a binding cavity suitable for the required ligand—however, the effect is localized and does not involve a large quantity of information. Similar considerations apply to proteins which form multichain complexes—in their case, information is encoded in the structure of their complexation interfaces. Of course, if a protein forms complexes with multiple other proteins, it must contain more information, i.e. a greater number of local deformations. In this case, information is encoded in hydrophobic residues exposed on the surface, where they come in contact with residues belonging to the intended complexation partner.

An even greater quantity of information is present in proteins which require the presence of a ligand (prosthetic group), and are also capable of forming complexes with other (potentially multiple) proteins, nucleic acids or ions. Notably, in the case of ions, complexation does not require deformations in the hydrophobic core structure since it is mediated by polar groups, which are typically exposed on the surface (naturally, there are exceptions: obviously the examples can be found—for example Zn-fingers where the presence of Zn ions is critical) The most complex protein structures in existence—sometimes referred to as biological machines (such as bacterial flagella) [1–7]—are beyond the scope of this study (the work on such objects is currently the object of our analysis); however, it can be safely assumed that their synthesis requires an even greater amount of information, due to their structural complexity and intricate mechanism of action.

The next group comprises proteins which call for the presence of a so-called permanent chaperone. The chaperone is a molecule which temporarily stabilizes a certain structural motif during the folding process, preventing undesirable conformational rearrangements which would result in a misfolded protein [8–10]. In some cases, biologically active proteins may require a permanent chaperone—for example, membrane proteins are only capable of performing their biological activity when attached to a membrane.

Membrane proteins are a large and diverse family. They usually lack a tertiary structure and cannot be likened to globular micelles—instead, they are embedded in cell membranes (or the surfaces of cellular organelles) and represent collections of secondary folds, such as helices or random coils. The membrane not only acts as a permanent chaperone, but in fact delivers other than water environment (other external force field). Removing this scaffold causes the membrane protein to undergo rearrangements which effectively wipe out its biological activity. Free from the stabilizing influence of the membrane, the protein refolds in a manner dictated by the properties of its environment—it may then be described as "biologically confused" and lacking a well-defined purpose. It turns into an ordinary chemical molecule and becomes a micelle—in this instance, a ribbonlike micelle, which is strongly deterministic and therefore lacks information.

Special object among proteins are amyloids. We treat them here as information-less systems. They appear to not carry any form of information, what can be recognized using the high symmetry criteria. All chains participating in amyloid repeat identical structure generating the ribbon-like micelle.

Experimental studies show that almost any protein may be converted into an amyloid by shaking [11–14]. In trying to explain this peculiar effect we note that shaking causes aeration of the solvent, which greatly increases its interphase (water/air) surface area. A critical property of the aqueous environment now comes into play: water is a fundamental requirement of life. This seemingly simple statement takes on a new meaning in the context of the presented phenomenon. Protein folding invariably occurs in an aqueous environment of which water is an ACTIVE participant. It generates an external force field which guides the folding process, favoring the formation of a micellar structure, which resembles a surfactant micelle. The fuzzy oil drop model acknowledges this effect by introducing a continuous force field, mathematically modeled as a 3D Gaussian. While the field acts upon the polypeptide chain, the chain is not always capable of "obeying" its instructions—simply because it lacks the structural freedom required to align itself with the field with perfect accuracy.

Local deviations from the theoretical hydrophobic core model may emerge in one of two ways:

1. because covalent bonds prevent the chain from reaching perfect (micelar) alignment,
2. as a result of external factors which stabilize certain deviations. Due to the targeted nature of this process (i.e. complexation of a specific ligand) in Ref. [15] we postulate that the ligand itself may be present during folding, and that its presence ensures the generation of a suitable binding cavity. This hypothesis has not been unequivocally validated, but it seems plausible in light of our discussion on the role of information in protein folding.

In subsequent chapters we will try to show that the aqueous environment—rather than being modeled as a collection of molecules—should instead be viewed as a continuous force field. Protein folding is not guided by individual molecules, and modeling atom-atom interactions between the protein and the solvent does not accurately capture the effect of the external force field upon such a large and complex structure as a polypeptide chain. Furthermore, our knowledge regarding the structural properties of water in its liquid phase is very limited. Why do biological processes call for a very specific concentration of NaCl in water (referred to as physiological or isotonic salinity)? It is likely that these conditions are necessary for the solvent to retain desirable structural properties, facilitating all processes associated with the machinery of life (except for some localized exceptions, such as low pH in the stomach).

Scientific research aimed at determining the structural properties of water (under both standard and altered physiochemical conditions) support the hypothesis that the aqueous environment plays an active role in biological processes. The phenomenon of amyloid transformation caused solely by physical factors (shaking), with no changes in the chemical structure of the solvent, further underscores the need for such research.

Thus far we have established the critical importance of water for protein folding. It is, however, equally interesting to speculate about the effect exerted by proteins upon the surrounding solvent.

The effect of proteins influence on the water structuralization

The active participation of water in protein folding process is expressed by directing the hydrophobic residues toward the center with simultaneous exposition o hydrophilic residues on the surface. Such interpretation is accordant with spontaneous process of micelle generation. On the other hand, the presence of objects like proteins is the source of information for surrounding water. Relatively large surface covered by charge atoms influence the ordering of water molecules in the close neighborhood. The good examples are the antifreeze proteins which alter the structural properties of the surrounding water so as to prevent formation of ice crystals. What tools does a protein have at its disposal in order to achieve this effect? While the properties of the solid state of water (ice) are reasonably well understood, the liquid phase still hides many unresolved questions—for example, it is not clear why the density of water peaks at 4 degrees Celsius, even though this phenomenon plays a critical role in enabling many biological processes.

The protein is able to affect its environment by exposing hydrophobicity and/or hydrophilicity on its surface. Even though—as remarked above—the structural properties of water in contact with various types of surfaces are poorly understood, we may assume that there are major differences in how water interacts with hydrophobic and hydrophilic patches exposed by the solvated molecule. These differences represent a mechanism by which the protein is able to transmit information to its environment.

In this context, it is especially interesting to consider the role of solenoids which are found in some antifreeze proteins. The solenoid is an axial structure, potentially capable of unlimited propagation along its principal axis. For this reason, solenoids used by antifreeze proteins are able to expose a unique type of surface, where alternating bands of high and low hydrophobicity strongly disrupt the natural structure of the surrounding solvent. Interestingly, some studies show that water molecules gain increased mobility when close to antifreeze proteins [16,17]. Clearly, this effect supports the intended function of antifreeze proteins, i.e. disrupting formation of ice crystals by altering the structural properties of water [18].

If solenoids represent a handy way to modulate the structure of the solvent, it might be interesting to search for them in other types of proteins. As it turns out, solenoid fragments are also commonly found in lyases. The difference, however, is that while in an antifreeze protein the sole purpose is to interact with the solvent, an enzyme plays a far more complex role. In addition to remaining soluble, it must also transmit a targeted signal to the intended substrate (i.e. it must be "recognized" by the substrate), and facilitate the intended reaction. The latter tasks requires that at least three distinct structural stages be available to the enzyme: precatalytic, intra-catalytic and post-catalytic. Each stage calls for a mechanism capable of enforcing conformational changes which lead to the following stage, while ensuring high specificity. At all stages, structural rearrangements (including reorganization of water ordering in the close neighborhood) must be subjected to tight control.

Solubility is achieved by creating a suitable "packaging." In enzymes (lyases), solenoids are surrounded by numerous random coil and helical fragments which appear to act as their "guardians." In most cases, these short fragments are well aligned with the Gaussian distribution and therefore promote solubility. Equally important are helices which exist at either end of the solenoid: these "caps" prevent the solenoid fragment from growing indefinitely (or forming a complex with another solenoid), which it would otherwise be capable of. Again, these helices are locally highly consistent with the 3D Gaussian distribution and admit water into close proximity of the protein, thus preventing complexation. The presence of helices along solenoid seems to play a role in mediation with water making the large molecule soluble and protected against uncontrolled complexation.

The structure of lyases is particularly intricate and it seems that their structural complexity acts as a "transmitter," sending out information to the environment. The signal alters the structural properties of water in a way which can be recognized by the intended ligand. Of course, water plays a key role at each step of this process—by reacting in specific ways to hydrophobic/hydrophilic conditions, it is capable of transmitting information across considerable distances. The same principles are employed in the so-called iceberg model of protein interaction [19].

In summary, the enzyme molecule is not only a tool which facilitates a certain reaction, but also a transmitter, which sends out signals by modulating the properties of its aqueous environment.

Amyloid structures

Where should amyloids be positioned in the presented hierarchy? As already mentioned, an amyloid may be characterized as a nonintelligent ribbonlike micelle, which—from the biological point of view—is generated for no reason, and fulfills no purpose. It is the result of physiochemical processes and carries no information. While their structure, consisting of alternating bands of high and low hydrophobicity, transmits a signal out into the environment, that signal also serves no purpose, other than altering the properties of the solvent. It is akin to transmitting and endless sequence of repeating tones. One may speculate that perhaps the signal is not recognized by any digestive enzyme and while amyloids exists untouched being resistant for degradation processes.

The protein is an intelligent micelle

In conclusion we can state that the protein is a highly intelligent micelle which carries information encoded in its own structure (proper fold), can send signals into the environment (e.g. to recognize its intended substrate or ligand), is capable of specific interaction, and has a specific activity profile, which may also account for its own degradation (interaction with digestive enzymes). While the aforementioned properties may seem difficult to describe using only 20 amino acids, it turns out that nature has managed to accomplish this goal.

If we accept the premise that proteins carry significant (though varied) amount of information, then their overall tendency to adopt micellar conformations (whether spherical or elongated) must be countered by a process which produces deviations from this theoretical model. While the micellar characteristics of proteins may be explained by invoking an external force field (i.e. a 3D Gaussian), it is not quite as obvious how certain targeted deviations emerge. One conclusion seems inescapable: the native form of a protein represents a compromise between holistic forces (the tendency to generate a micelle) and specific means of encoding information. Models which focus only on optimization of internal force fields cannot accurately capture this property.

When submitting this work for publication the authors remarked that, unlike other similar books, this one is prospective in character. To-date publications which the authors have had the opportunity to study adopt a retrospective approach to amyloidosis: they refer to past experiments and entrenched theories. In contrast, the authors prospectively suggest that further research on amyloids should acknowledge the structural properties of the aqueous environment. These properties—which include, among others, on the presence of solvated ions, nonpolar compounds, temperature and aeration—may, in some cases, promote generation of structures capable of unrestricted elongation, and devoid of "caps" (unlike solenoid fragments found in many biologically active proteins). In general terms, understanding the ways in which proteins control and exploit information (e.g. to interact with other molecules) is necessary to explain certain biological processes on the molecular level—including amyloidosis, which should be regarded as a chemical phenomenon rather than a biological one, and therefore not subject to laws which control the behavior of living organisms.

The protein may be compared to an intelligent micelle, while an amyloid represent a nonintelligent ribbonlike micelle. The amyloid transformation process robs the protein of information describing its intended role, and instead produces a repetitive, deterministic structure consisting of multiple identical units (peptides). This repetitive character explains the bandlike distribution of hydrophobicity in amyloid fibrils. Amyloidogenesis (which often depends on external factors, such as shaking) means introducing order among fragments with identical sequences—especially from the point of view of hydrophobicity distribution. A distribution which diverges from the theoretical monocentric Gaussian—for example by introducing a hydrophilic breaker between two local hydrophobicity maxima—may represent a seed for amyloid transformation. This is evident e.g. in the V domain of the immunoglobulin light chain.

In light of the hypothesis presented in this chapter, the process of amyloidogenesis corresponds to forfeiture of information, which produces an information-free structure. The process strips the biological molecule of its purpose-built nature and turns it into a nonintelligent entity, subject only to physiochemical mechanisms. This is the way—as speculated—of transthyretin amyloid transformation (Chapter 12).

The authors are aware that the presented view is controversial. Nevertheless, the ubiquitous presence of water and its effect on polypeptide chains seem to be of key importance in explaining the properties of proteins. If this controversial hypothesis engenders further debate, the authors will consider their work to have been a success. Likewise, we would be very grateful for any critical remarks our readers would care to formulate.

As shown in Fig. 4.1 amyloid transformation may occur spontaneously (*in vivo*), which is considered a pathological process. *In vitro* almost any protein can be converted into an amyloid, as indicated by arrows in the figure.

The presence and role or water environment is discussed in other papers [20,21]. Characteristics and physico-chemical properties of water particularly water in contact with hydrophobic surface are discussed in details in Refs. [22–28].

The list of references given below represents the step-wise development of the model presented here and they can be treated as the base for hypotheses presented in this work [29–43].

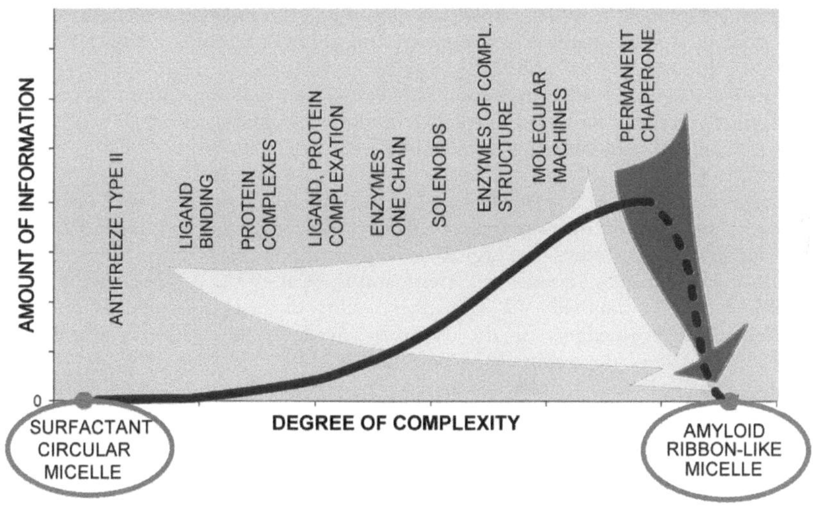

Fig. 4.1 Graphical depiction of the increased quantity of information (arbitrary units) carried by increasingly complex protein structures. Spherical and ribbonlike micelles are regarded carrying low information quantity due to their high degree of deterministic symmetry. Yellow arrow (white in print version)—amyloid transformation as observed *in vitro* (by shaking); red arrow (gray in print version)—spontaneous amyloid transformation *in vivo*, mostly affecting proteins which lose their permanent chaperone and adopt deterministic forms which lead to amyloid structures. Teal frame and dot—zero-information state. Gray frames—single-chain molecules. Dashed line—pathological process.

References

[1] Streif S, Staudinger WF, Marwan W, Oesterhelt D. Flagellar rotation in the Archaeon *Halobacterium salinarum* depends on ATP. Journal of Molecular Biology 2008;384(1): 1–8. https://doi.org/10.1016/j.jmb.2008.08.057.

[2] Silverman M, Simon M. Flagellar rotation and the mechanism of bacterial motility. Nature 1974;249(5452):73–4. https://doi.org/10.1038/249073a0.

[3] Kasai T, Sowa Y. Measurements of the rotation of the flagellar motor by bead assay. The Bacterial Flagellum 2017:185–92. https://doi.org/10.1007/978-1-4939-6927-2_14.

[4] Khan S, Jain S, Reid GP, Trentham DR. The fast tumble signal in bacterial chemotaxis. Biophysical Journal 2004;86(6):4049–58. https://doi.org/10.1529/biophysj.103.033043.

[5] Berg HC, Anderson RA. Bacteria swim by rotating their flagellar filaments. Nature 1973;245(5425):380–2. https://doi.org/10.1038/245380a0.

[6] Jahn TL, Bovee EC. Movement and locomotion of microorganisms. Annual Review of Microbiology 1965;19(1):21–58. https://doi.org/10.1146/annurev.mi.19.100165.000321.

[7] Harshey RM. Bacterial motility on a surface: many ways to a common goal. Annual Review of Microbiology 2003;57(1):249–73. https://doi.org/10.1146/annurev.micro.57.030502.091014.

[8] Selkoe DJ. erratum: folding proteins in fatal ways. Nature 2004;428(6981). https://doi.org/10.1038/nature02477. 445–445.

[9] Dobson CM. Protein folding and misfolding. Nature 2003;426(6968):884–90. https://doi.org/10.1038/nature02261.

[10] Soto C, Estrada L, Castilla J. Amyloids, prions and the inherent infectious nature of misfolded protein aggregates. Trends in Biochemical Sciences 2006;31(3):150–5. https://doi.org/10.1016/j.tibs.2006.01.002.

[11] Cui D, Ou S, Patel S. Protein-spanning water networks and implications for prediction of protein-protein interactions mediated through hydrophobic effects. Proteins: Structure, Function, and Bioinformatics 2014;82(12):3312–26. https://doi.org/10.1002/prot.24683.

[12] Ladner-Keay CL, Griffith BJ, Wishart DS. Shaking Alone induces De Novo conversion of recombinant prion proteins to β-sheet rich oligomers and fibrils. PLoS One 2014;9(6):e98753. https://doi.org/10.1371/journal.pone.0098753.

[13] Cohen SIA, Linse S, Luheshi LM, Hellstrand E, White DA, Rajah L, Knowles TPJ. Proliferation of amyloid- 42 aggregates occurs through a secondary nucleation mechanism. Proceedings of the National Academy of Sciences 2013;110(24): 9758–63. https://doi.org/10.1073/pnas.1218402110.

[14] Giehm L, Otzen DE. Strategies to increase the reproducibility of protein fibrillization in plate reader assays. Analytical Biochemistry 2010;400(2):270–81. https://doi.org/10.1016/j.ab.2010.02.001.

[15] Nielsen L, Khurana R, Coats A, Frokjaer S, Brange J, Vyas S, Fink AL. Effect of environmental factors on the kinetics of insulin fibril formation: elucidation of the molecular mechanism†. Biochemistry 2001;40(20):6036–46. https://doi.org/10.1021/bi002555c.

[16] Konieczny L, Roterman-Konieczna I. Conclusion. Protein folding in silico. 2012. p. 191–202. https://doi.org/10.1533/9781908818256.191.

[17] Laage D. A molecular jump mechanism of water reorientation. Science 2006; 311(5762):832–5. https://doi.org/10.1126/science.1122154.

[18] Modig K, Qvist J, Marshall CB, Davies PL, Halle B. High water mobility on the ice-binding surface of a hyperactive antifreeze protein. Physical Chemistry Chemical Physics 2010;12(35):10189. https://doi.org/10.1039/c002970j.

[19] Banach M, Konieczny L, Roterman I. Why do antifreeze proteins require a solenoid? Biochimie 2018;144:74–84. https://doi.org/10.1016/j.biochi.2017.10.011.
[20] Ben-Naim A. Myths and verities in protein folding theories: from Frank and Evans iceberg-conjecture to explanation of the hydrophobic effect. The Journal of Chemical Physics 2013;139(16):165105. https://doi.org/10.1063/1.4827086.
[21] Serpell LC. Alzheimer's amyloid fibrils: structure and assembly. Biochimica et Biophysica Acta (BBA) - Molecular Basis of Disease 2000;1502(1):16–30. https://doi.org/10.1016/s0925-4439(00)00029-6.
[22] Al-Garawi ZS, Morris KL, Marshall KE, Eichler J, Serpell LC. The diversity and utility of amyloid fibrils formed by short amyloidogenic peptides. Interface Focus 2017;7(6): 20170027. https://doi.org/10.1098/rsfs.2017.0027.
[23] Kim KH, Späh A, Pathak H, Perakis F, Mariedahl D, Amann-Winkel K, Nilsson A. Maxima in the thermodynamic response and correlation functions of deeply supercooled water. Science 2017;358(6370):1589–93. https://doi.org/10.1126/science.aap8269.
[24] Gallo P, Stanley HE. Supercooled water reveals its secrets. Science 2017;358(6370): 1543–4. https://doi.org/10.1126/science.aar3575.
[25] Palmer JC, Martelli F, Liu Y, Car R, Panagiotopoulos AZ, Debenedetti PG. Metastable liquid–liquid transition in a molecular model of water. Nature 2014; 510(7505):385–8. https://doi.org/10.1038/nature13405.
[26] Tanaka H. A self-consistent phase diagram for supercooled water. Nature 1996; 380(6572):328–30. https://doi.org/10.1038/380328a0.
[27] Macias-Romero C, Nahalka I, Okur HI, Roke S. Optical imaging of surface chemistry and dynamics in confinement. Science 2017;357(6353):784–8. https://doi.org/10.1126/science.aal4346.
[28] Schutzius TM, Jung S, Maitra T, Graeber G, Köhme M, Poulikakos D. Spontaneous droplet trampolining on rigid superhydrophobic surfaces. Nature 2015;527(7576): 82–5. https://doi.org/10.1038/nature15738.
[29] Roterman-Konieczna I. Protein folding in silico. 2012. https://doi.org/10.1533/9781908818256.
[30] Jurkowski W, Baster Z, Dułak D, Roterman-Konieczna I. The early-stage intermediate. Protein Folding in Silico 2012:1–20. https://doi.org/10.1533/9781908818256.1.
[31] Banach M, Konieczny L, Roterman-Konieczna I. The late-stage intermediate. Protein Folding in Silico 2012:21–37. https://doi.org/10.1533/9781908818256.21.
[32] Alejster P, Jurkowski W, Roterman-Konieczna I. Structural information involved in the interpretation of the stepwise protein folding process. Protein Folding in Silico 2012:39–54. https://doi.org/10.1533/9781908818256.39.
[33] Banach M, Marchewka D, Piwowar M, Roterman-Konieczna I. The divergence entropy characterizing the internal force field in proteins. Protein Folding in Silico 2012: 55–77. https://doi.org/10.1533/9781908818256.55.
[34] Banach M, Konieczny L, Roterman-Konieczna I. Ligand-binding-site recognition. Protein Folding in Silico 2012:79–93. https://doi.org/10.1533/9781908818256.79.
[35] Banach M, Konieczny L, Roterman-Konieczna I. Use of the "fuzzy oil drop" model to identify the complexation area in protein homodimers. Protein Folding in Silico 2012:95–122. https://doi.org/10.1533/9781908818256.95.
[36] Konieczny L, Roterman-Konieczna I. Simulation of the polypeptide chain folding process using the "fuzzy oil drop" model. Protein Folding in Silico 2012:123–40. https://doi.org/10.1533/9781908818256.123.
[37] Roterman I, Banach M, Konieczny L. Application of the fuzzy oil drop model describes amyloid as a ribbonlike micelle. Entropy 2017;19(4):167. https://doi.org/10.3390/e19040167.

[38] Roterman I, Banach M, Kalinowska B, Konieczny L. Influence of the aqueous environment on protein structure—a plausible hypothesis concerning the mechanism of amyloidogenesis. Entropy 2016;18(10):351. https://doi.org/10.3390/e18100351.

[39] Banach M, Kalinowska B, Konieczny L, Roterman I. Role of Disulfide bonds in stabilizing the conformation of selected enzymes—an approach based on divergence entropy Applied to the structure of hydrophobic core in proteins. Entropy 2016;18(3):67. https://doi.org/10.3390/e18030067.

[40] Jurkowski W, Brylinski M, Konieczny L, Wiíniowski Z, Roterman I. Conformational subspace in simulation of early-stage protein folding. Proteins: Structure, Function, and Bioinformatics 2004;55(1):115–27. https://doi.org/10.1002/prot.20002.

[41] Brylinski M, Jurkowski W, Konieczny L, Roterman I. Limited conformational space for early-stage protein folding simulation. Bioinformatics 2004;20(2):199–205. https://doi.org/10.1093/bioinformatics/btg391.

[42] Dułak D, Gadzała M, Banach M, Ptak M, Wiśniowski Z, Konieczny L, Roterman I. Filamentous aggregates of tau proteins fulfil standard amyloid criteria provided by the fuzzy oil drop (FOD) model. International Journal of Molecular Sciences 2018;19(10):2910. https://doi.org/10.3390/ijms19102910.

[43] Dułak D, Banach M, Gadzała M, Konieczny L, Roterman I. Structural analysis of the Aβ(15-40) amyloid fibril based on hydrophobicity distribution. Acta Biochimica Polonica 2018. https://doi.org/10.18388/abp.2018_2647.

CHAPTER 5

Proteins structured as spherical micelles

Mateusz Banach[1], Leszek Konieczny[2], Irena Roterman[1]
[1]Department of Bioinformatics and Telemedicine, Jagiellonian University–Medical College, Krakow, Poland
[2]Chair of Medical Biochemistry, Jagiellonian University–Medical College, Krakow, Poland

Contents

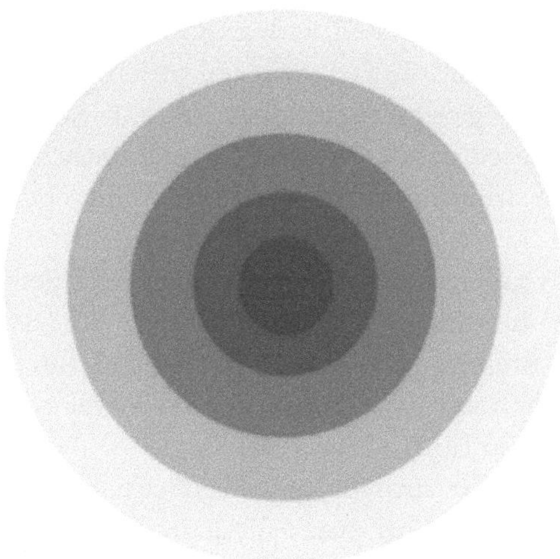

The image depicts a highly regular, spherically symmetrical structure. Shades of gray correspond to increasing concentrations of hydrophobicity, which is low on the

From Globular Proteins to Amyloids
ISBN: 978-0-08-102981-7
https://doi.org/10.1016/B978-0-08-102981-7.00005-1

surface but high at the center. The only difference between the presented image and the theoretical Gaussian lies in the fact that the Gaussian distribution is continuous rather than discrete.

The model presented in this work, referred to as the fuzzy oil drop model (FOD), is supported by the existence of proteins which conform to it with near-perfect accuracy. Such proteins exhibit excellent agreement between the observed distribution of hydrophobicity (abbreviated as O) and the corresponding theoretical distribution (abbreviated as T) which is expressed as a 3D Gaussian. Accordingly, these proteins may be thought of as spherical micelles.

The image attached to the title of this chapter visualizes a highly ordered structure which exhibits spherical symmetry. Shades of gray correspond to increasing concentrations of hydrophobicity, which is low on the surface but peaks at the center. The only difference between the presented image and the theoretical Gaussian lies in the fact that the Gaussian distribution is continuous rather than discrete.

Proteins rated as highly consistent with the theoretical model include type III antifreeze proteins, muscle tissue proteins such as titin, as well as certain fast-folding proteins. In all these cases tertiary conformations are stabilized by prominent hydrophobic cores rather than by other factors (i.e. disulfide bonds).

With regard to type III antifreeze proteins, their conformance to the fuzzy oil drop model is associated with exposure of polar residues on the surface. This has an effect on the structural ordering of water molecules, however the resulting structuralization of the solvent differs from conditions encountered in ice crystals. Consequently, type III antifreeze proteins lower the freezing point of water simply by virtue of being present.

Muscle tissue is subject to frequent stretching and deformations caused by external forces. In such cases it is desirable for its constituent proteins to revert to their original form when no external forces are present. A well-ordered hydrophobic core provides this useful reversibility, which may be observed i.e. in titin.

Finally, the group of strongly accordant structures includes proteins referred to as "fast-folding" (or even "ultrafast-folding"). In their case, the missing information required at the folding stage is clearly provided by the aqueous solvent.

Type III antifreeze proteins

This group is represented by an antifreeze protein isolated from ocean pout, *Macrozoarces americanus*, listed in PDB under ID 1AME [1]. It consists of 66 residues arranged into four helices and four β-strands forming two anti-parallel strands. In addition, it also contains short disordered fragments.

Fig. 5.1 illustrates the theoretical and observed distribution of hydrophobicity in antifreeze protein type III (1AME).

For this protein, RD values for T-O-R and T-O-H are equal to 0.300 and 0.294 respectively. Correlation coefficients are as follows: HvT = 0.370; TvO = 0.796; HvO = 0.707. These figures indicate the presence of a prominent hydrophobic core (RD < 0.5), and the protein is therefore classified as accordant with the model. In addition, high values of both TvO and HvO suggest that the protein's tendency to form a near-perfect spherical micelle is consistent with the intrinsic properties of its individual residues (notwithstanding limitations described in the introduction, where micellar protein structures are compared to surfactant

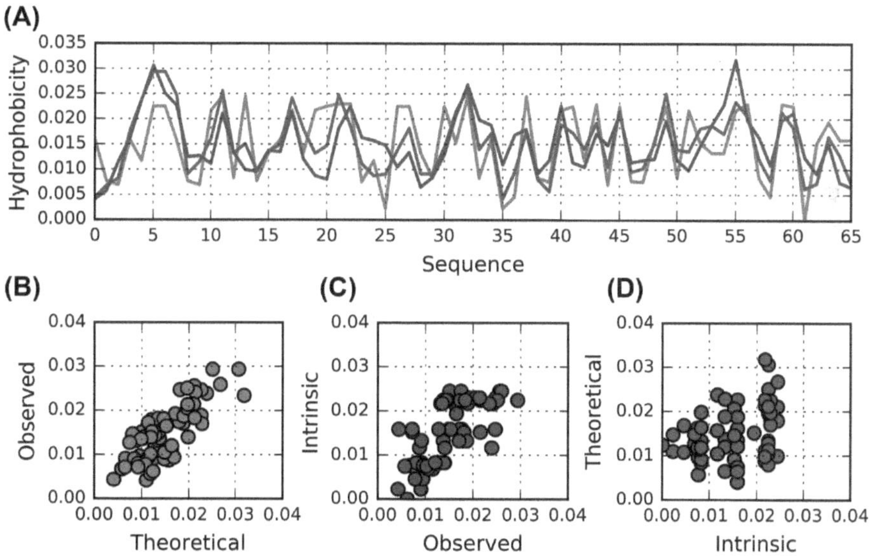

Fig. 5.1 FOD characteristic of antifreeze protein type III (1AME): (A) hydrophobicity distribution profiles: T (blue (dark gray in print version)), O (red (gray in print version)) and H (green (light gray in print version)), (B), (C), (D) correlation scatter plots (TvO, HvO, and HvT respectively). Each circle in (B), (C) and (D) represents one residue in the sequence.

micelles). Note that a low value of HvT would suggest that the theoretical and intrinsic distributions contradict each other. The presented protein serves as proof that, in spite of problems caused by covalent bonds between residues, as well as variable hydrophobicity of the residues themselves, it is nevertheless possible for the protein to achieve a fold which closely corresponds to the spherical model. Good agreement between T and O indicates that the protein's surface consists of polar residues while hydrophobic residues congregate in its central part.

Fig. 5.2 provides view of structure of good accordance with the fuzzy oil drop model. The biological role of this protein is to induce a specific type of order in the surrounding water (though exposure of polarity), hindering the formation of ice crystals. The antifreeze properties of 1AME are thus explained.

Later on we will discuss other possible ways of counteracting crystallization of water (see chapter 7).

Titin

Titin is primarily found in muscle tissue. It is a giant protein, consisting of a sequence of β-sandwich domains. Its overall length may reach 1 μm [2]. Titin works by mimicking a coil spring: it deforms under the influence of external forces, and reverts to its original shape in the absence of such forces.

The immunoglobulin-like domain found in titin differs from immunoglobulins in that it lacks disulfide bonds. Under such conditions the only factor capable of stabilizing the protein's tertiary conformation is a well-ordered hydrophobic core. The core enables the domain to revert to its original structure when no deforming forces are present.

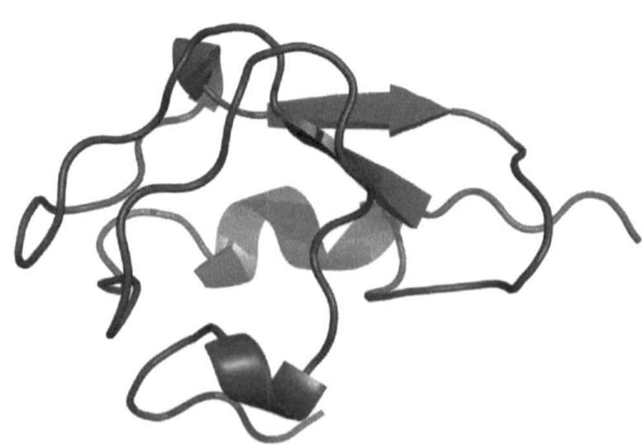

Fig. 5.2 3D presentation of antifreeze protein type III (1AME).

Since the structure of titin (PDB ID: 1TIT) was thoroughly described in another publication [3], here we will focus on the so-called human titin-obscurin-1 complex (PDB ID: 2WWM) [4], which consists of two titin domains and two obscurin-like domains (referred here as obscurin for brevity). Published research underscores the medical importance of this complex since point mutations affecting its sequence may results in hereditary muscle diseases [5].

From the point of view of our current subject it is important that each domain participating in the complex retains a highly ordered hydrophobic core.

As noted, the 2WWM complex consists of four chains labeled O, C (obscurin domains) and T, D (titin domains) respectively. Both proteins are described as "mainly β-sandwich" (2.60.40.10) in the CATH classification. Their length is also comparable: 100 aa for titin and 98 aa for the obscurin domain.

Due to the chain-like arrangement of T/O and D/C dimers which are separated by sparsely packed areas, there is no reason to expect that the four-chain complex as a whole will contain a well-ordered hydrophobic core. Our analysis therefore focuses on individual domains. The status of each chain from titin and obscurin is listed in Table 5.1 and Fig. 5.3. Low RD values for both T-O-R and T-O-H) indicate that the observed distribution of hydrophobicity is, in each case, closely aligned with the theoretical distribution and therefore the domain contains a prominent hydrophobic core. Correlation coefficients appear balanced, with no distinctive outliers. This suggests synergistic self-ordering of residues in the process of shaping the core of each domain.

Table 5.1 Comparison of fuzzy oil drop parameters, showing that each domain contains a prominent hydrophobic core but the complex as a whole does not contain a shared core as observed in titin-obscurin (2WWM).

	RD		CC		
Chain	**T-O-R**	**T-O-H**	**HvT**	**TvO**	**HvO**
Complex	0.768				
Chain C	0.442	0.458	0.729	0.572	0.437
Chain D	0.391	0.345	0.440	0.671	0.755
Chain O	0.464	0.492	0.413	0.530	0.758
Chain T	0.395	0.359	0.479	0.644	0.793

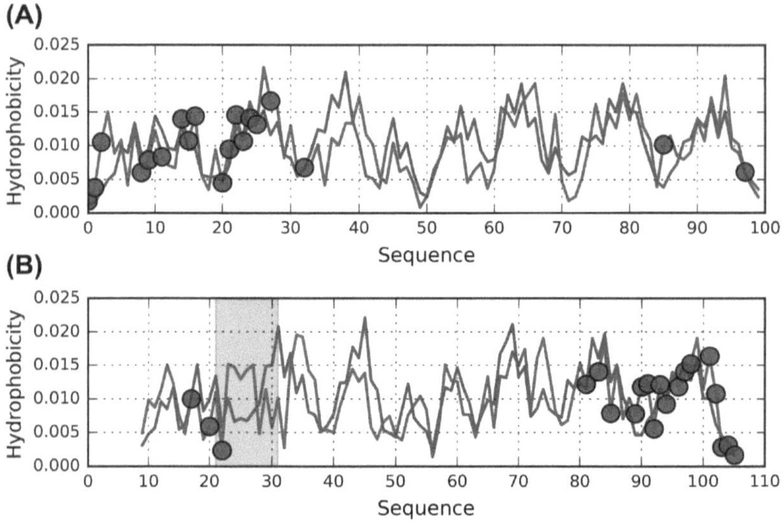

Fig. 5.3 Theoretical (T, blue (dark gray in print version)) and observed (O, red (gray in print version)) hydrophobicity distribution profiles for titin-obscurin complex (2WWM): (A) titin (chain D). (B) obscurin (chain O).

Magenta circles denote residues engaged in protein-protein interactions between the chains. The sole local discordance between the hydrophobicity distributions in obscurin (highlighted in red) does not cause the entire domain to be regarded as discordant with the FOD model.

Hydrophobicity distribution profiles (Fig. 5.3) also reveal good agreement between T and O, suggesting the presence of a strong, central hydrophobic core. In effect, both domains should be regarded as highly accordant with the model and structurally stabilized by their respective hydrophobic cores (Fig. 5.4.).

Such strong accordance with the monocentric structural pattern is also related to the protein's biological purpose. As a component of muscle tissue, it is frequently subjected to external forces and must revert to its original conformation in their absence. Given that no disulfide bonds are present, tertiary conformational stability is mediated solely by the hydrophobic core, and such core must be prominent enough to exert a powerful effect upon the protein's structure. An additional argument which supports our hypothesis regarding the purpose of titin's well-structured hydrophobic core relates to the use of this molecule in protein unfolding analyzes [6]. Studies which focus on (possibly multistage) unfolding often measure the elongation of a protein as a function of the applied longitudinal force. In

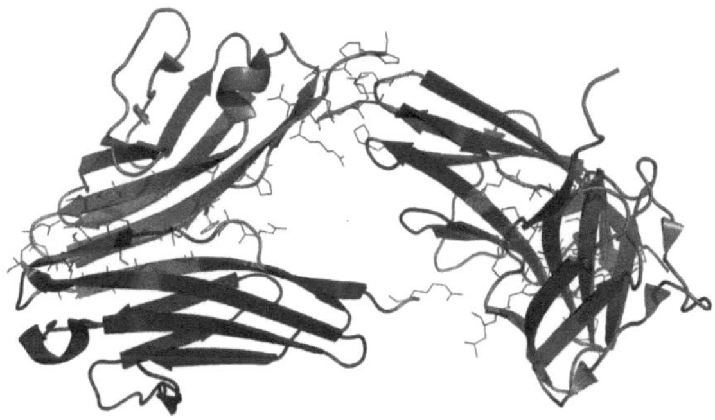

Fig. 5.4 3D presentation of titin (blue (dark gray in print version)) and obscurin (marine (gray in print version)) domains in titin-obscurin complex (2WWM). The fragments marked in each chain of obscurin in red (21–31) are locally discordant. Residues engaged in protein-protein interactions are magenta-colored and have side chains displayed.

order to achieve a macroscopic effect, titin is often used as a convenient scaffold for the target protein. Notably, titin itself undergoes single-step unfolding, thus any elongation becomes easy to detect and identify. This phenomenon appears to relate to the properties of titin's hydrophobic core which undergoes rapid destruction when the applied force exceeds a certain threshold.

An in-depth comparative analysis of β-sandwich domains, particularly in the context of immunoglobulins and their constituent parts, can be found [3]. In this chapter we focus on identification of prominent hydrophobic cores in muscle tissue proteins.

Ultrafast-folding proteins

This group of proteins is represented by mutant chicken villin subdomain hp-35 (PDB ID: 2F4K) with two point mutations: K65L and K70L. Swapping lysine for leucine at both positions produces a protein referred to as fast-folding, or even ultrafast-folding [7]. Given the strong polarity of K and the equally strong hydrophobicity of L, this change (as seen in Fig. 5.5) greatly strengthens the protein's hydrophobic core. The corresponding RD values are: 0.319 (T-O-R) and 0.204 (T-O-H), with correlation coefficient 0.468, 0.764 and 0.735 (HvT, TvO and HvO respectively). Notably, these are among the lowest RD values we have ever encountered throughout our years of work with the fuzzy oil drop model.

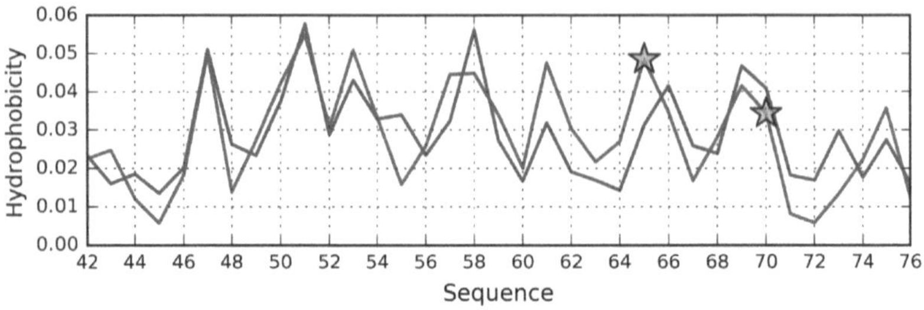

Fig. 5.5 Theoretical (T, blue (dark gray in print version)) and observed (O, red (gray in print version)) hydrophobicity distribution profiles for Chicken villin subdomain (2F4K). Orange stars mark the mutation loci (K65L and K70L).

Taken together, the above values indicate the presence of a particularly prominent monocentric hydrophobic core. Excellent alignment between O and T (regardless of which other reference distribution—R or H—is selected) and balanced correlation coefficients with notable domination of TvO confirm strong micellar properties of this molecule. This observation is further supported by plotting theoretical and observed hydrophobicity distributions, as shown in Fig. 5.5.

Visual inspection of Fig. 5.5 leads to unequivocal conclusion—both distributions are very similar. The role of both previously mentioned mutations is also visualized: introduction of two hydrophobic residues (L) in place of hydrophilic residues (K) at positions where the theoretical model expects hydrophobicity to remain high (in the center of the molecule, as shown in Fig. 5.6) promotes the formation of a hydrophobic core and may explain the extremely rapid ("ultrafast") nature of this process.

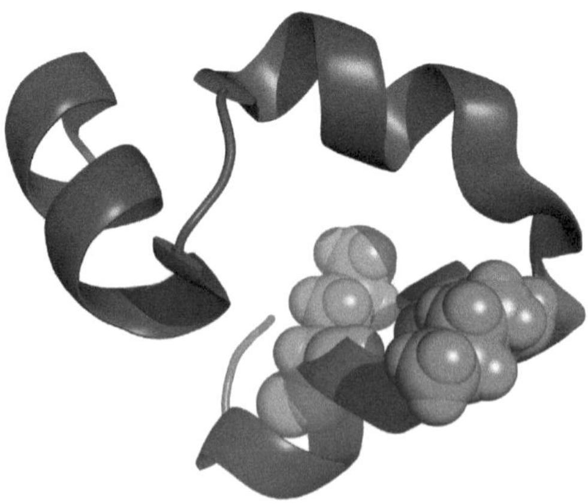

Fig. 5.6 3D presentation of chicken villin subdomain (2F4K). Orange spheres (gray in print version) mark the mutation loci (K65L and K70L).

It might be interesting to consider the presented case in a broader context. Such considerations would touch upon the status of the aqueous solvent which appears to play an active role in the folding process, guiding its course and determining the formation of a hydrophobic core—all in line with the basic assumptions of the fuzzy oil drop model. In Ref. [7] the authors state the following: "The ultrafast folding rate, very accurate X-ray structure, and small size make this engineered villin subdomain an ideal system for simulation by atomistic molecular dynamics with explicit solvent" (excerpt from the abstract). Clearly, the presented protein represents a convenient study subject, enabling validation of the model itself.

Screening analysis of a comprehensive non-redundant set of proteins [8] confirms that the vast majority of domains folds in accordance with the fuzzy oil drop model. This serves as further proof of the model's correctness and applicability in simulating various secondary and supersecondary folds [9].

Regarding immunoglobulin-like domains, despite strong topological similarities their constituent β-strands tend to exhibit variable accordance with the model [3].

The presented protein chicken villin subdomain (2F4K) contains no disulfides. This suggests that its structural stability is solely due to hydrophobic interactions. Calculation of fuzzy oil drop parameters appears to confirm this view. Another protein stabilized by its hydrophobic core will be presented in the chapter which discusses the relation between structural stability and disulphide bonds (Chapter 7) as well as in Refs. [10,11]. As it turns out, both factors, i.e. hydrophobic interactions and disulfides, may either reinforce or counteract each other and therefore either promote or frustrate tertiary stability.

Stabilization in the presence of disulfide bonds

Modern biochemistry textbooks single out hydrophobic cores and disulfide bonds as factors which stabilize the tertiary conformations of proteins. In the example presented in this chapter both factors act cooperatively and are mutually reinforcing. Our analysis concerns a bioactive peptide named τ-AnmTx Ueq 12-1 (short name: Ueq 12-1), isolated from the sea anemone *Urticina eque* listed in PDB under ID 5LAH—a short protein (45 aa) which nevertheless contains five disulfide bonds [10].

Bioactive peptide—Ueq 12-1 (5LAH) is a highly peculiar protein owing to the arrangement of its cysteines. Three adjacent Cys residues form three separate disulfide bonds—a rarely observed pattern. The multitude of disulfides renders the protein very stable; what is more, the hydrophobic core is also extraordinarily well ordered and consistent with the theoretical distribution of hydrophobicity (Table 5.2)—this goes for individual secondary folds as well as for fragments connecting Cys residues which form disulfide bonds. To enable comparative analysis we also list the corresponding correlation coefficients.

Taken together, the presented parameters reveal high consistency between T and O, which, in turn, shows that the protein adopts a micelle-like conformation. The network of disulfide bonds appears to further reinforce this structure. Both factors (hydrophobicity and disulfide bonds) work toward the same goal. Detailed analysis of individual folds shows that each of them is also consistent with the theoretical model (Fig. 5.7, Fig. 5.8).

Table 5.2 Values of fuzzy oil drop parameters calculated for the structure of 5LAH.

	RD		**Correlation coefficients**		
Bioactive peptide (5LAH)	**T-O-R**	**T-O-H**	**HvT**	**TvO**	**HvO**
PROTEIN	0.294	0.321			
SS					
1-8	0.228	0.309	0.798	0.787	0.668
11-42	0.306	0.348	0.432	0.764	0.744
17-35	0.353	0.321	0.309	0.702	0.688
22-43	0.318	0.450	0.447	0.754	0.793
29-44	0.296	0.549	0.550	0.755	0.869
β-structure					
9-11	0.141	0.119	0.549	0.996	0.478
32-34	0.184	0.104	0.671	0.962	0.847
β-sheet	0.178	0.095	0.350	0.926	0.600
β-structure					
16-18	0.446	0.063	0.889	0.988	0.810
41-44	0.490	0.973	0.310	0.337	0.987
27-29	0.373	0.734	0.493	0.703	0.965
β-sheet	0.422	0.577	0.424	0.562	0.844

"SS" corresponds to fragments bracketed by disulfide bonds. Individual β-strands and the resulting β-sheets are also listed.

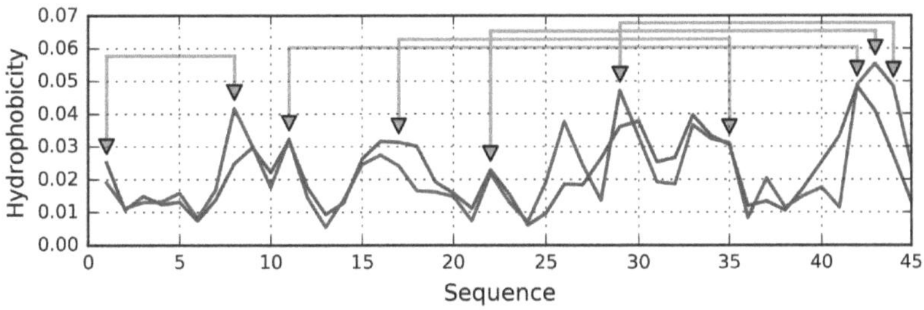

Fig. 5.7 Theoretical (T, blue (dark gray in print version)) and observed (O, red (gray in print version)) hydrophobicity distribution profiles for bioactive peptide Ueq 12-1 (5LAH). Orange lines correspond to SS bonds formed by Cys residues.

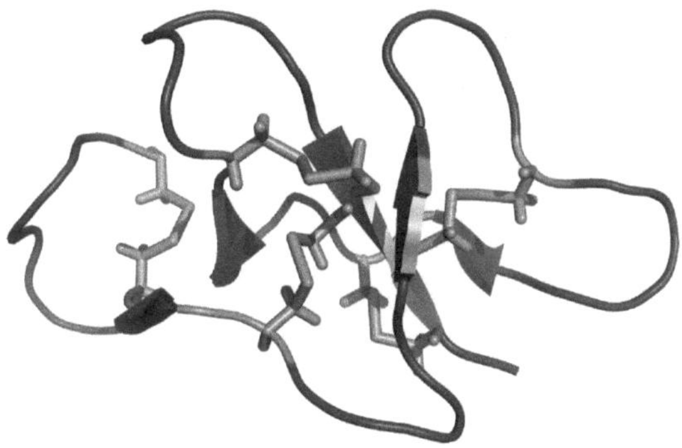

Fig. 5.8 3D presentation of bioactive peptide Ueq 12-1 (5LAH) showing the location of disulfides.

Proteins consistent with the theoretical model—spherical micelles

Referring back to the hypothesis stated in the introduction, that the presence of a hydrophobic core is a result of interactions between the peptide and its aqueous environment, we propose that the protein resembles a micelle whose structure is determined by hydrophobic interactions (notwithstanding differences between the building blocks of proteins and surfactant micelles). The final product—when characterized by a monocentric distribution of hydrophobicity—may be likened to a spherical micelle. In this section we discuss proteins which, regardless of structural factors not found in surfactant micelles (variable intrinsic hydrophobicity of amino acids; covalent bonding between components), adopt a spherical micelle-like conformation. Ultrafast-folding proteins provide a particularly fitting example of this group.

With regard to our previous discussion concerning sources of information required in the folding process, the presented proteins show that—in addition to information carried by the peptide sequence itself—the process also relies on information obtained from the aqueous solvent, which directs polar residues to the surface while shielding hydrophobic residues from contact with water. As can be seen on the example of 5LAH, the system of disulfide bonds may remain in perfect harmony with the hydrophobic core structure—although there are also proteins where such bonds prevent an organized hydrophobic core from forming (an example is discussed in Ref. [11]).

For the sake of contrast we will now discuss a protein which strongly diverges from the expected distribution of hydrophobicity.

Counter-example—prefoldin

The prefoldin β-subunit from *Thermococcus strain KS-1* (PDB ID: 2ZQM) exhibits major deviations from the expected distribution of hydrophobicity. It consists of two anti-parallel helices linked by a protruding β-hairpin. While it essentially lacks a tertiary conformation, CATH categorizes is as 1.10.287.370—mainly alpha orthogonal bundle [12].

Fuzzy oil drop model parameters calculated for this protein are as follows: RD(T-O-R) = 0.621; RD(T-O-H) = 0.619; HvT = 0.035; TvO = 0.249; HvO = 0.715. Taken together, these values indicate that the protein's structure is dominated by the conformational preferences of individual residues and that it lacks a monocentric hydrophobic core (note the low value of TvO and the correspondingly high value of HvO).

Fig. 5.9 illustrates the scope of discordance between T and O. Together with Fig. 5.10 it proves that micellar characteristics are by no means universal among proteins.

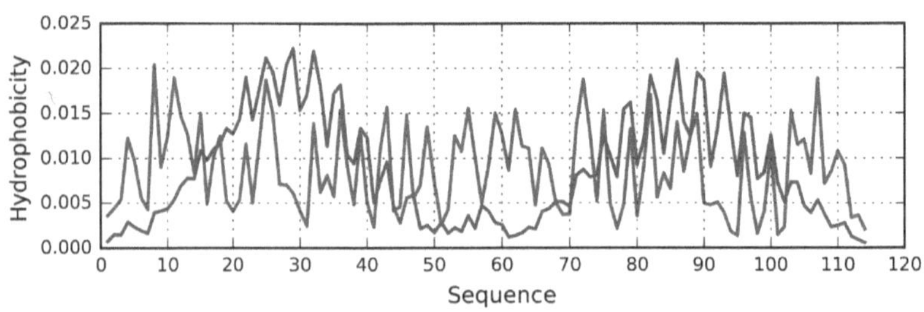

Fig. 5.9 Theoretical (T, blue (dark gray in print version)) and observed (O, red (gray in print version)) hydrophobicity distribution profiles for prefoldin β-subunit (2ZQM) revealing protein's major deviations from the expected distribution of hydrophobicity.

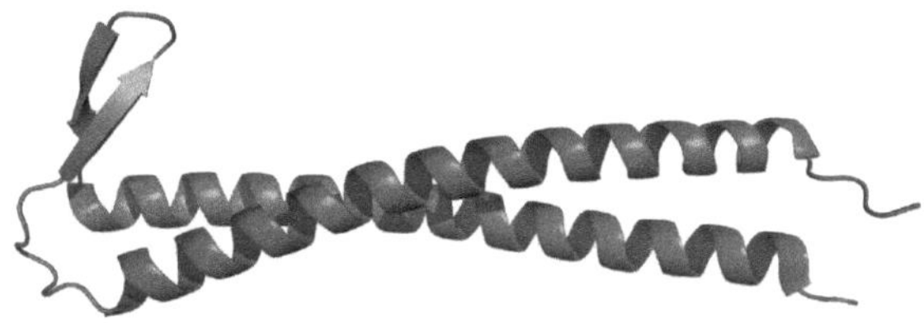

Fig. 5.10 3D presentation of prefoldin β-subunit (2ZQM).

Small proteins in particular (<60 aa) exhibit a whole range of discordances between O and T. Nevertheless, the presented work aims on validating the core assumptions of the fuzzy oil drop model, which is why we focus on proteins (largely) accordant with the model.

References

[1] Ye Q, Leinala E, Jia Z. Structure of type III antifreeze protein at 277 K. Acta Crystallographica Section D Biological Crystallography 1998;54(4):700–2. https://doi.org/10.1107/s0907444997020040.

[2] Labeit S, Barlow DP, Gautel M, Gibson T, Holt J, Hsieh C-L, Trinick J. A regular pattern of two types of 100-residue motif in the sequence of titin. Nature 1990; 345(6272):273–6. https://doi.org/10.1038/345273a0.

[3] Banach M, Konieczny L, Roterman I. The fuzzy oil drop model, based on hydrophobicity density distribution, generalizes the influence of water environment on protein structure and function. Journal of Theoretical Biology 2014;359:6–17. https://doi.org/10.1016/j.jtbi.2014.05.007.

[4] Pernigo S, Fukuzawa A, Bertz M, Holt M, Rief M, Steiner RA, Gautel M. Structural insight into M-band assembly and mechanics from the titin-obscurin-like-1 complex. Proceedings of the National Academy of Sciences 2010;107(7):2908–13. https://doi.org/10.1073/pnas.0913736107.

[5] Göktepe S, Abilez OJ, Parker KK, Kuhl E. A multiscale model for eccentric and concentric cardiac growth through sarcomerogenesis. Journal of Theoretical Biology 2010;265(3):433–42. https://doi.org/10.1016/j.jtbi.2010.04.023.

[6] Munoz V, Cerminara M. When fast is better: protein folding fundamentals and mechanisms from ultrafast approaches. Biochemical Journal 2016;473(17):2545–59. https://doi.org/10.1042/bcj20160107.

[7] Kubelka J, Chiu TK, Davies DR, Eaton WA, Hofrichter J. Sub-microsecond protein folding. Journal of Molecular Biology 2006;359(3):546–53. https://doi.org/10.1016/j.jmb.2006.03.034.

[8] Sałapa K, Kalinowska B, Jadczyk T, Roterman I. Measurement of hydrophobicity distribution in proteins - non-redundant protein data bank. Bio-Algorithms and Med-Systems 2012;8. https://doi.org/10.2478/bams-2012-0023.

[9] Kalinowska B, Banach M, Wiśniowski Z, Konieczny L, Roterman I. Is the hydrophobic core a universal structural element in proteins? Journal of Molecular Modeling 2017;23(7). https://doi.org/10.1007/s00894-017-3367-z.

[10] Logashina YA, Solstad RG, Mineev KS, Korolkova YV, Mosharova IV, Dyachenko IA, Andreev YA. New disulfide-stabilized fold provides sea anemone peptide to exhibit both antimicrobial and TRPA1 potentiating properties. Toxins 2017; 9(5):154. https://doi.org/10.3390/toxins9050154.

[11] Banach M, Kalinowska B, Konieczny L, Roterman I. Role of disulfide bonds in stabilizing the conformation of selected enzymes—an approach based on divergence entropy applied to the structure of hydrophobic core in proteins. Entropy 2016;18(3):67. https://doi.org/10.3390/e18030067.
[12] Kida H, Sugano Y, Iizuka R, Fujihashi M, Yohda M, Miki K. Structural and molecular characterization of the prefoldin β subunit from Thermococcus strain KS-1. Journal of Molecular Biology 2008;383(3):465–74. https://doi.org/10.1016/j.jmb.2008.08.041.

CHAPTER 6

Local discordance

Mateusz Banach[1], Leszek Konieczny[2], Irena Roterman[1]
[1]Department of Bioinformatics and Telemedicine, Jagiellonian University – Medical College, Krakow, Poland
[2]Chair of Medical Biochemistry, Jagiellonian University – Medical College, Krakow, Poland

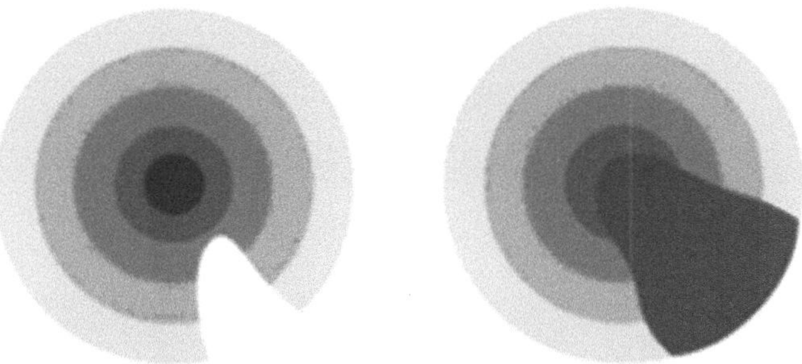

The images visualizes a single-chain proteins with two forms of discordance present: local deficiency and local excess. The first one assumed to be ready to interact with other molecules (including substrate) and to be ready for complexation using the local hydrophobicity excess on the surface as the target area for protein-protein complexation.

Local discordance is understood as local hydrophobicity deficiency or local hydrophobicity excess. The first one appears to be related to the presence of cavity. The second one—particularly when the exposure takes place on the surface—may suggest potential protein-protein complexation area.

Presence of cavity allows substrate binding—as it is shown for enzymes, as well as the permanent ligand binding. The examples for both cases are shown in this chapter.

From Globular Proteins to Amyloids
ISBN: 978-0-08-102981-7
https://doi.org/10.1016/B978-0-08-102981-7.00006-3

CHAPTER 6.A

The active site in a single-chain enzyme

Mateusz Banach[1], Leszek Konieczny[2], Irena Roterman[1]
[1]Department of Bioinformatics and Telemedicine, Jagiellonian University – Medical College, Krakow, Poland
[2]Chair of Medical Biochemistry, Jagiellonian University – Medical College, Krakow, Poland

Contents

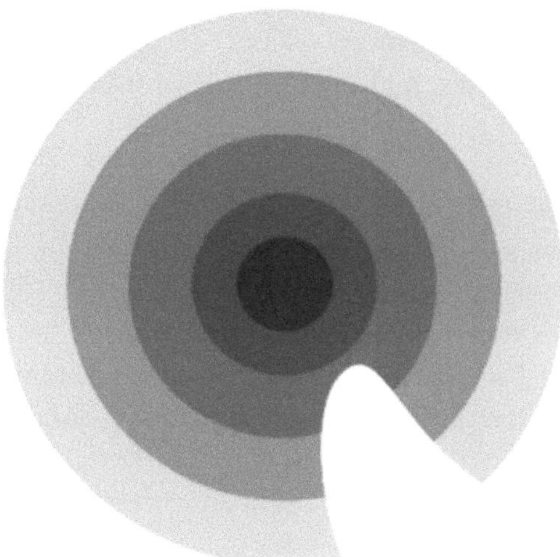

The image visualizes a single-chain enzyme which, while largely accordant with the 3D Gaussian distribution, also contains a binding cavity. In terms of the fuzzy oil drop model the cavity manifests itself as a localized hydrophobicity deficiency. It encodes information which determines the function of the protein.

From Globular Proteins to Amyloids
ISBN: 978-0-08-102981-7
https://doi.org/10.1016/B978-0-08-102981-7.00007-5

A single-chain enzyme should—in accordance with information theory—encode information enabling it to recognize the intended substrate and perform catalysis. Its structure should be shaped by the aqueous environment (i.e. it should resemble a spherical micelle) and feature a local deficit of hydrophobicity in the area of its binding cavity. The deficit expresses information which determines the enzyme's biological role. We will investigate this hypothesis by considering two distinct single-chain enzymes: a lysozyme and a ribonuclease.

In this work the hydrolase family is represented by human lysozyme (EC 3.2.1.17, PDB ID: 1:Z1) [1] and *Bos taurus* pancreatic ribonuclease (EC.3.1.27.5, PDB ID: 5RSA) [2]. Both have a diverse secondary structure with multiple disulphide bonds and a clearly distinguished catalytic active site.

Lysozyme

Table 6.A.1 lists the structural properties of the lysozyme (PDB ID: 1LZ1). In light of the presented results, the lysozyme can be said to deviate from the theoretical distribution of hydrophobicity. Its RD (T-O-R) is greater than 0.5, with balanced values of all correlation coefficients (although HvO dominates, high TvO indicates strong involvement of hydrophobic forces in shaping the protein's structure).

Regarding catalytic residues and their immediate neighborhood, these fragments are also identified as discordant. Eliminating these residues lowers the RD(T-O-R) value for the remainder of the protein.

Fig. 6.A.1. Theoretical (T, blue) and observed (O, red) hydrophobicity distribution profiles for 1LZ1. Magenta background denotes helical fragment while yellow background denotes β-strands. Orange stars mark the catalytic residues. Orange lines correspond to SS-bonds formed by Cys residues.

The structure of the lysozyme is further stabilized by four disulfide bonds. Fragments bracketed by these bonds vary in terms of their FOD status: 6—128 and 30—116 are both identified as discordant while the other two fragments remain consistent with the model. Notably, the discordant fragments (6—128 and 30—116) contain both of the enzyme's catalytic residues. It is therefore possible that their discordance is not directly caused by the presence of disulfides.

When analyzing individual secondary folds, we note that the β-sheet as a whole, as well as all its component β-strands, significantly diverge from the

Table 6.A.1 Values of fuzzy oil drop parameters calculated for the structure of 1LZ1 and its selected fragments. Values printed in boldface reflect discordance. The leftmost column lists secondary folds and (where applicable) the location of catalytic residues.

Lysozyme (1LZ1)		RD		Correlation coefficient		
Fragment	**AA**	**T-O-R**	**T-O-H**	**HvT**	**TvO**	**HvO**
Complete mol.	1–130	**0.529**	0.477	0.387	0.515	0.781
Catalytic active site						
Cat. Res. absent	35, 53	**0.515**	0.465	0.432	0.547	0.779
35E Close neighborhood ±5 residues	30–41	**0.504**	0.170	0.088	0.393	0.701
53D Close neighborhood ±5 residues	48–59	**0.545**	**0.649**	0.541	0.687	0.791
SS-bonds						
SS-bonds	6–128	**0.531**	**0.504**	0.400	0.507	0.789
35E, 53D	30–116	**0.543**	0.462	0.387	0.499	0.762
	65–81	0.279	0.279	0.608	0.836	0.822
	77–95	0.488	**0.734**	0.365	0.493	0.832
Secondary structure						
β-strands	42–46	**0.620**	**0.750**	-0.194	0.345	0.525
53D	51–55	**0.603**	**0.626**	0.088	0.421	0.854
	59–61	**0.858**	0.428	0.802	-0.416	0.207
β-sheet		**0.617**	**0.514**	0.292	0.442	0.462
Helices	4–15	0.376	0.398	0.573	0.661	0.912
35E	24–37	**0.565**	0.213	0.190	0.237	0.642
	81–86	0.443	**0.779**	0.533	0.458	0.881
	89–101	0.263	0.118	0.562	0.778	0.813
	104–109	**0.502**	**0.612**	0.896	0.440	0.767
	109–116	0.317	**0.626**	0.637	0.696	0.899
	121–125	0.129	**0.622**	0.920	0.966	0.961

theoretical distribution of hydrophobicity. This is an important observation given the presence of the catalytic residue at 53D. Likewise, the helical fold which contains the enzyme's other catalytic residue (35E) is also discordant.

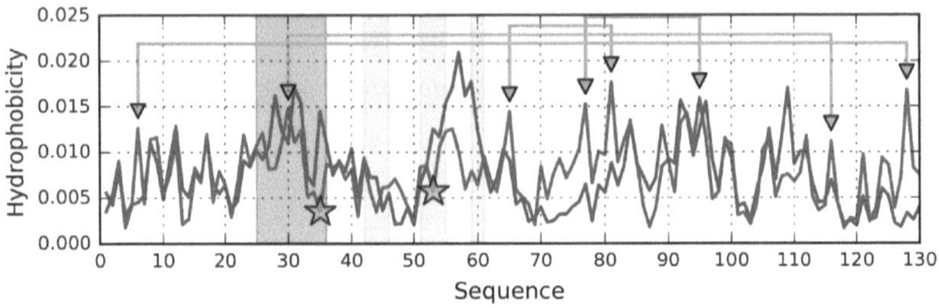

Fig. 6.A.1 Illustrates the status of the lysozyme, highlighting fragments which are identified as discordant in Table 6.A.1. As shown, all these fragments are close to the active site.

By invoking our formal hypothesis, which conflates information content with departures from the deterministic distribution of hydrophobicity, we may claim that the β-sheet and the neighboring helix carry information required by the enzyme to fulfill its role.

Fig. 6.A.2 clearly shows an outer shell which surrounds the active site. This shell exhibits micellar properties and is generated via interaction with the aqueous solvent. As already noted, information content determines activity—this includes the capability to recognize the substrate, as well as a blueprint for conformational changes required in the process of catalysis (which is inherently dynamic). It seems that the latter property may be linked to the presence of a centrally placed helix which plays an important part in stabilizing the molecule as a whole.

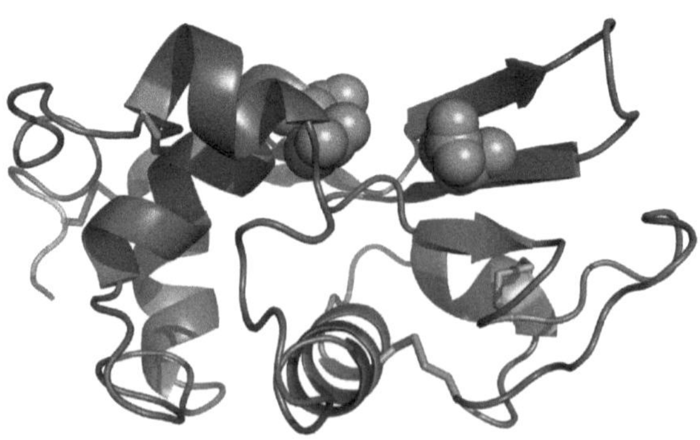

Fig. 6.A.2 3D presentation of 1LZ1. Fragments highlighted in red are identified by fuzzy oil drop model as discordant (helix: 24–37, sheet: 42–46, 51–55, 59–61). Orange spheres mark the catalytic residues (53D and 35E). Orange sticks correspond to disulfides.

Referring again to the status of the β-sheet, the status of its individual folds varies. The fragment at 51–55 is significantly affected by intrinsic hydrophobicity and strongly discordant—as evidenced by its high value of RD. The fragment at 42–46 is characterized by negative correlation coefficients coupled with high values of RD—we may describe it as being "in active opposition" to the theoretical distribution. The fragment therefore contributes a large quantity of information to the β-sheet and the entire neighborhood of the active site.

Ribonuclease

Ribonuclease (PDB ID: 5RSA) provides another example of how information can be encoded in a single-chain enzyme. Similarly to the lysozyme, we expect the protein to include a region characterized by strong discordance from the theoretical distribution, reflecting its high information content (Fig. 6.A.3 and Fig. 6.A.4).

It should be noted that fragments which comprise and surround the catalytic active site are non-micellar in character. These fragments are believed to encode information which determines the enzyme's specific activity profile.

As can be seen in Table 6.A.2, the discordance of the additional white fragment (87–90) identified in the presented hydrophobicity profiles (Fig. 6.A.4) may be related to the flexibility of the outlying loop which provides a way for the substrate to migrate to its required location.

The structure of ribonuclease exhibits major deviations from the theoretical hydrophobicity profile in areas which comprise its active site. Much like

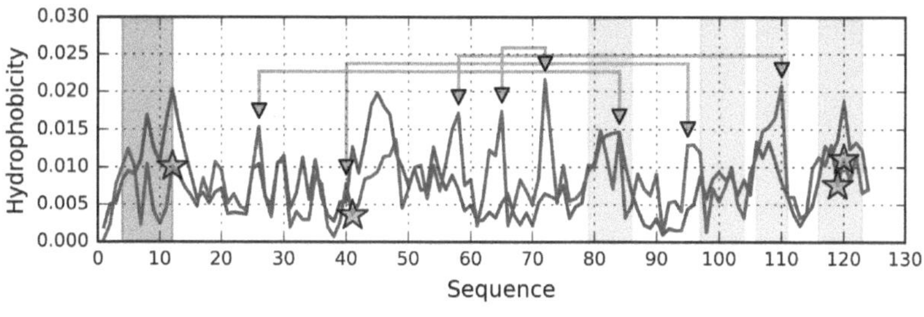

Fig. 6.A.3 Theoretical (T, blue) and observed (O, red) hydrophobicity distribution profiles for 5RSA. Magenta background denotes helical fragment while yellow background denotes β-sheets. Orange stars mark the catalytic residues. Orange lines correspond to SS bonds formed by Cys residues.

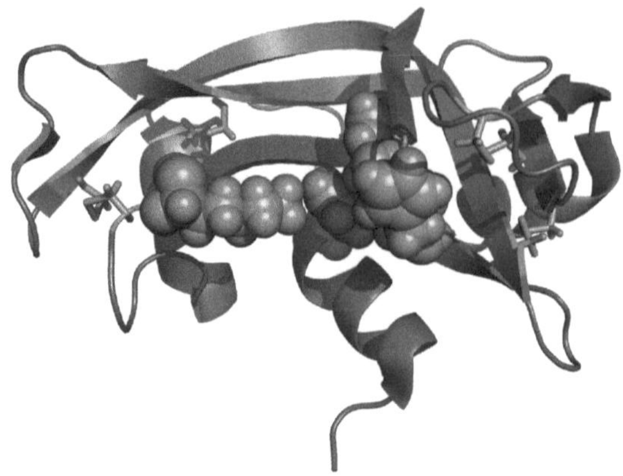

Fig. 6.A.4 3D presentation of ribonuclease (5RSA). Fragments highlighted in red are identified by fuzzy oil drop model as discordant (helix: 24–37, sheet: 42–46, 51–55, 59–61). Orange spheres mark the catalytic residues (53D and 35E). Orange sticks correspond to disulfides.

Table 6.A.2 Values of fuzzy oil drop parameters calculated for the structure of ribonuclease (5RSA) and its selected fragments. Values printed in boldface reflect discordance expressed either by high RD values or biased domination of HvO correlation coefficient. The leftmost column lists secondary folds and (where applicable) the location of catalytic residues. Underscored values are regarded as particularly important for determination of conformational characteristics.

Ribonuclease (5RSA)		RD		Correlation coefficient		
Fragment	AA	T-O-R	T-O-H	HvT	TvO	HvO
Complete mol.	1–124	**0.550**	**0.563**	0.300	0.378	0.815
Catalytic center						
Close neighborhood ±5 residues						
12H	7–17	0.469	0.315	0.385	0.548	0.837
41K	36–46	0.407	0.281	0.278	0.647	0.820
119H, 120F	114–124	**0.602**	0.411	0.084	0.472	0.442
SS-bonds						
SS-bonds fragments	26–84	**0.534**	**0.551**	0.308	0.378	0.836
	40–95	**0.633**	**0.611**	0.205	0.312	0.807
	58–110	**0.536**	**0.565**	0.304	0.431	0.848
	65–72	0.437	**0.563**	0.165	0.473	0.916

Table 6.A.2 Values of fuzzy oil drop parameters calculated for the structure of ribonuclease (5RSA) and its selected fragments. Values printed in boldface reflect discordance expressed either by high RD values or biased domination of HvO correlation coefficient. The leftmost column lists secondary folds and (where applicable) the location of catalytic residues. Underscored values are regarded as particularly important for determination of conformational characteristics.—cont'd

Secondary structure						
β-strands	42–48	0.443	0.381	0.200	0.512	0.572
	79–87	**0.542**	0.530	0.389	0.457	0.878
	96–105	**0.548**	0.359	**0.087**	**-0.013**	**0.862**
β-sheet		**0.557**	0.443	**0.386**	**0.378**	**0.811**
β-strands	61–63	0.342	0.163	0.870	0.947	0.981
	71–75	0.459	**0.863**	0.289	0.349	0.980
	106–111	**0.575**	0.487	0.568	0.168	0.792
119H, 120F	**116–124**	**0.657**	0.280	-0.093	0.228	0.276
β-sheet		**0.591**	**0.530**	0.337	0.227	0.738
Helices 12H	**3–13**	**0.568**	0.477	**0.249**	**0.304**	**0.887**
	24–33	0.320	0.503	0.763	0.707	0.939
	50–56	0.477	0.272	0.514	0.706	0.891
	57–60	0.201	0.412	0.758	0.828	0.945

in the case of the lysozyme, this micellar "capsule" protects the active site, which—in all likelihood—must remain discordant to ensure a suitable environment for catalysis.

In summary, we may note that each of the presented enzymes includes fragments which resemble an organized micelle. In these fragments the observed distribution of hydrophobicity is consistent with the theoretical Gaussian, and they may be regarded as a spherical (or globular) micelle.

In contrast, the fragments marked in red exhibit major deviations from the theoretical distribution of hydrophobicity. By actively opposing micellization, such fragments carry information which the protein requires to fulfill its biological purpose.

Unlike surfactant micelles which consist of identical unit molecules and retain perfect symmetry, the protein may be described as an "intelligent micelle". Local deviations from symmetric patterns play an important role in this context.

FOD-based folding simulations carried out for these two proteins are described in Refs. [3,4].

References

[1] Artymiuk PJ, Blake CCF. Refinement of human lysozyme at 1.5 Å resolution analysis of non-bonded and hydrogen-bond interactions. Journal of Molecular Biology 1981; 152(4):737–62. https://doi.org/10.1016/0022-2836(81)90125-x.

[2] Wlodawer A, Borkakoti N, Moss DS, Howlin B. Comparison of two independently refined models of ribonuclease-A. Acta Crystallographica Section B Structural Science 1986;42(4):379–87. https://doi.org/10.1107/s0108768186098063.

[3] Jurkowski W, Bryliński M, Konieczny L, Roterman I. Lysozyme Folded In Silico According to the limited conformational sub-space. Journal of Biomolecular Structure and Dynamics 2004;22(2):149–57. https://doi.org/10.1080/07391102.2004.10506991.

[4] Jurkowski W, Brylinski M, Konieczny L, Wiśniowski Z, Roterman I. Conformational subspace in simulation of early-stage protein folding. Proteins: Structure, Function, and Bioinformatics 2004;55(1):115–27. https://doi.org/10.1002/prot.20002.

CHAPTER 6.B

Protein-protein interaction encoded as an exposure of hydrophobic residues on the surface

Mateusz Banach[1], Leszek Konieczny[2], Irena Roterman[1]
[1]Department of Bioinformatics and Telemedicine, Jagiellonian University—Medical College, Krakow, Poland
[2]Chair of Medical Biochemistry, Jagiellonian University—Medical College, Krakow, Poland

Contents

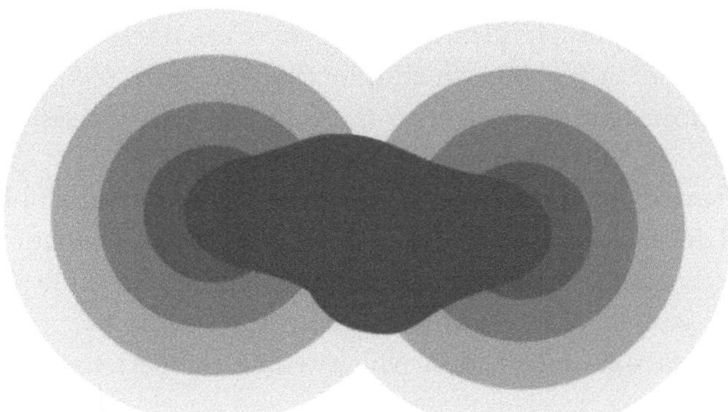

According to the fuzzy oil drop model, information required for complexation of an external polypeptide chain should be encoded as a local exposure of hydrophobicity on the protein surface. Of course, this is not the sole mechanism by which protein complexes may emerge; however, exposed hydrophobicity appears to play an important part in the complexation process. A local excess of hydrophobicity — especially if

From Globular Proteins to Amyloids
ISBN: 978-0-08-102981-7
https://doi.org/10.1016/B978-0-08-102981-7.00008-7

occurring on the surface — creates an environment which favors interaction with other polypeptide chains; particularly those, which also expose a hydrophobic interface.

This chapter discusses a homodimer which meets all the above conditions: DNA-binding regulatory protein — *Escherichia coli* met repressor in complex with a co-repressor (S-adenosylmethionine) and DNA (PDB ID: 1CMA) [1], as well as in the absence of these molecules (PDB ID: 1CMB) [2], i.e. in the apo form.

To reveal the role of the complexation interface we calculate model parameters for the complex, for individual chains treated as components of the complex and for standalone individual chains (adjusting the shape of the 3D Gaussian capsule as necessary). In addition, we determine the status of interface residues in the complex as well as in individual chains.

As shown in Table 6.B.1, the apo dimer is consistent with the model, while the corresponding structure, which includes the co-repressor and the DNA fragment, exhibits deviations from the model. A common property of the apo dimer and the larger complex is that their interfaces are FOD-accordant (in the area of inter-chain interactions) while the remainder of the structure exhibits elevated RD values, which indicates that the interface plays an important role in stabilizing the complex as a whole. When deprived of interface residues, the dimers exhibit poor structural stability.

As a whole, the apo dimer is accordant, while the co-repressor complex is not. When attempting to identify the causes of this discrepancy, we should note the fragment at 88—99, which is locally discordant (i.e. eliminating it from calculation significantly lowers the RD value for the remaining part of the chain).

When treated as a standalone structure, each monomer exhibits characteristic local deviations between O and T. In all cases these deviations may be attributed to residues which comprise the interface. Clearly, the structure of the interface is entropically disadvantageous in a monomer — indeed, its purpose is to encode information which would facilitate the formation of a dimer.

The above hypothesis may be criticized by noting that we are analyzing each chain as it appears in the dimer, and that, when analyzed on its own, it might not exhibit such an anomalous distribution of hydrophobicity. This can be countered by noting that even when chains undergo structural realignment during dimerization, they must first possess the capacity for such realignment, and that the information required in this process must be encoded in their structure.

Table 6.B.1 Status of the apo dimer (1CMB) and the dimer in complex with a co-repressor (1CMA), computed for the whole dimer, for individual chains treated as part of the dimer and for standalone individual chains respectively. The table also lists the corresponding values which result from elimination of interface residues (no P—P) and for the interface residues themselves (P—P), as well as elimination of other selected fragments (to identify the micellar part of the protein). Values listed in boldface indicate discordance.

Complex in apo form (1CMB)

		RD		Correlation coefficient		
Apo complex (1CMB)	**Fragment**	**T-O-R**	**T-O—H**	**HvT**	**TvO**	**HvO**
Complex		0.489	0.402	0.371	0.616	0.766
	Chain A	0.449	0.351	0.378	0.649	0.757
	Chain B	**0.522**	0.446	0.365	0.585	0.776
Complex						
No P—P		**0.554**	0.450	0.108	0.412	0.734
P—P		0.380	0.214	0.465	0.700	0.728
Chain A						
No P—P		**0.531**	0.406	0.038	0.400	0.712
P—P		0.336	0.173	0.520	0.760	0.718
Chain B						
No P—P		**0.570**	0.487	0.170	0.427	0.753
P—P		0.419	0.257	0.401	0.630	0.738

(Continued)

Table 6.B.1 Status of the apo dimer (1CMB) and the dimer in complex with a co-repressor (1CMA), computed for the whole dimer, for individual chains treated as part of the dimer and for standalone individual chains respectively. The table also lists the corresponding values which result from elimination of interface residues (no P—P) and for the interface residues themselves (P—P), as well as elimination of other selected fragments (to identify the micellar part of the protein). Values listed in boldface indicate discordance.—cont'd

Complex in apo form (1CMB)						
		RD		Correlation coefficient		
Apo complex (1CMB)	**Fragment**	**T-O-R**	**T-O—H**	**HvT**	**TvO**	**HvO**
Chain A – individual						
Chain A		**0.568**	0.453	0.148	0.415	0.720
No P—P		**0.530**	0.405	0.073	0.352	0.714
P—P		**0.658**	**0.561**	0.265	0.558	0.698
DNA binding		0.473	0.182	0.174	0.762	0.625
No DNA bin.		**0.571**	0.479	0.152	0.395	0.708
Eliminated residues	45-47, 95-100, 68—72, 79—83	0.453	0.297	0.221	0.441	0.711
Chain B – individual						
Chain B		**0.586**	0.488	0.137	0.367	0.730
No P—P		**0.539**	0.450	0.134	0.356	0.750
P—P		**0.701**	**0.603**	0.134	0.417	0.692

Complex including co-repressor (1CMA)						
1CMA	Fragment	RD		Correlation coefficient		
Complex incl. co-repressor		T-O-R	T-O–H	HvT	TvO	HvO
Complex		**0.518**	0.428	0.344	0.570	0.763
	Chain A	**0.520**	0.437	0.346	0.560	0.757
	Chain B	**0.516**	0.420	0.343	0.581	0.770
Eliminated residues	88–99	0.388	0.310	0.400	0.689	0.750
Chain A + Chain B – status in complex						
No P–P		**0.578**	0.447	0.164	0.410	0.726
P–P		0.423	0.318	0.441	0.615	0.770
Chain A		**0.520**	0.437	0.346	0.560	0.757
No P–P		**0.578**	0.454	0.144	0.399	0.714
P–P		0.436	0.333	0.450	0.590	0.763
Chain B		**0.516**	0.420	0.343	0.581	0.770
No P–P		**0.578**	0.440	0.183	0.421	0.739
P–P		0.410	0.302	0.432	0.641	0.778
Chain A – individual						
Chain A		**0.592**	0.485	0.149	0.359	0.716
No P–P		**0.557**	0.432	0.178	0.349	0.713
P–P		**0.688**	**0.671**	0.143	0.440	0.709
Eliminated residues	P–P and 3–5,88-99	0.481	0.311	0.260	0.555	0.657

(Continued)

Table 6.B.1 Status of the apo dimer (1CMB) and the dimer in complex with a co-repressor (1CMA), computed for the whole dimer, for individual chains treated as part of the dimer and for standalone individual chains respectively. The table also lists the corresponding values which result from elimination of interface residues (no P—P) and for the interface residues themselves (P—P), as well as elimination of other selected fragments (to identify the micellar part of the protein). Values listed in boldface indicate discordance.—cont'd

Complex in apo form (1CMB)						
		RD		**Correlation coefficient**		
Apo complex (1CMB)	**Fragment**	**T-O-R**	**T-O—H**	**HvT**	**TvO**	**HvO**
Chain B – individual						
Chain B		**0.587**	**0.645**	0.134	0.356	0.728
No P—P		**0.561**	0.421	0.195	0.333	0.736
P—P		**0.658**	**0.619**	0.040	0.476	0.683
Eliminated residues	P—P and 3—5, 33, 88-99	0.476	0.294	0.318	0.557	0.687

It is reasonable to conclude that a hydrophobic area — especially when located on the molecular surface — may mediate complexation with external proteins. Of course, this is not the only means of identifying complexation interfaces. The CAPRI project [3] shows that it is possible to pinpoint interfaces by studying nonbonding interactions and geometric properties of monomers; nevertheless, the presented line of reasoning, based on the fuzzy oil drop model, may provide useful results at least for a subset of protein complexes.

A detailed discussion of how various types of homodimers are formed can be found in Ref. [4], which distinguishes several types of relations between monomers depending on FOD accordance/discordance of the resulting dimer. This study proposes a new dimerization mechanism which is not strictly based on geometric alignment, but instead relies on the presence of a "quasi-domain" in the interface area, to which both monomers contribute, and which remains highly consistent with the fuzzy oil drop model. This domain is believed to structurally stabilize proteins which must withstand external forces — for example dystrophin, which connects the cytoskeleton to muscular proteins and is therefore subject to frequent stretching.

Table 6.B.1 reveals that, in some cases, eliminating interface residues does not cause RD to become lower than 0.5. It might be interesting to speculate about other possible causes of this phenomenon. As it turns out, the fragment at 88—99 is particularly discordant in nearly all presented structures (especially when a co-repressor is involved). As shown in the 3D presentation, this discordance may be attributed to the exposed outlying loop, which is structurally flexible and may undergo frequent conformational changes.

Residues involved in dimerization, distinguished in Fig. 6.B.1, are typically characterized by excess hydrophobicity. This is particularly true of the fragment at 20—40. Other fragments which diverge from the theoretical distribution (marked in Fig. 6.B.1) are either too hydrophobic or insufficiently hydrophobic given their location. Eliminating them from calculations lowers the corresponding RD values (Figs. 6.B.2 and Fig. 6.B.3).

Using a homodimer as a representative example shows how local deviations from the theoretical distribution of hydrophobicity may assist dimerization. Given that the interface is highly accordant in the dimer, we may conclude that — in the scope of an individual chain — it represents a targeted form of discordance, specifically constructed to enable interactions with the intended complexation partner (Fig. 6.B.4)

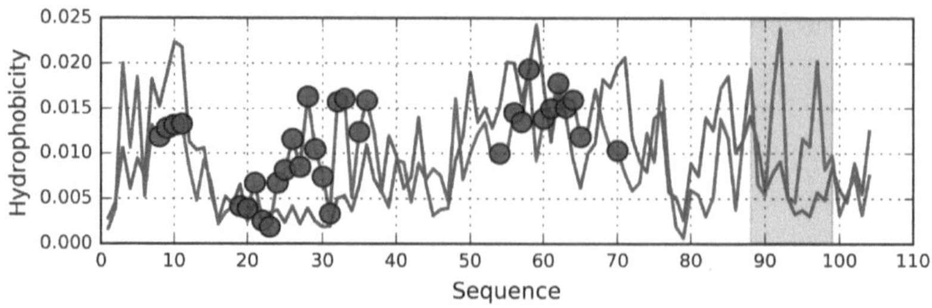

Fig. 6.B.1 Theoretical (T, blue) and observed (O, red) hydrophobicity distribution profiles for 1CMB chain A. Magenta circles mark residues involved in P-P interaction. Red background indicates a discordant fragment (88–99), which must be eliminated along with interacting residues to produce a value of RD below 0.5.

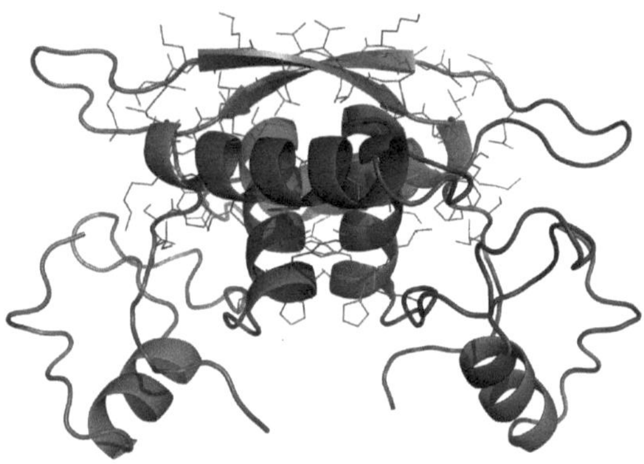

Fig. 6.B.2 3D presentation of 1CMB: chain (A) blue, chain (B) gray. Residues engaged in protein-protein interactions are magenta-colored and have side chains displayed. The fragments marked in each chain in red (88–99) are locally discordant.

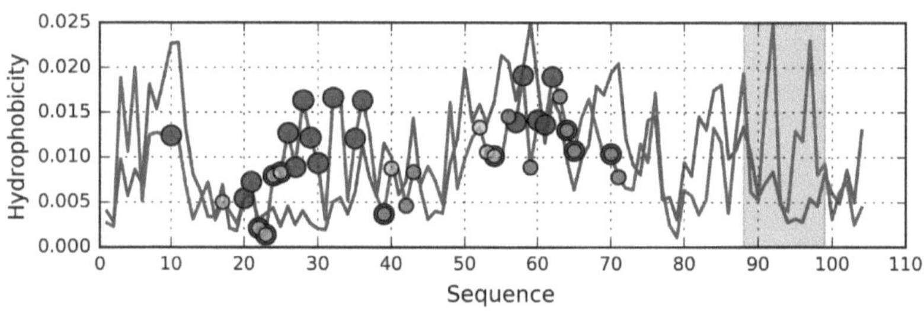

Fig. 6.B.3 Theoretical (T, blue) and observed (O, red) hydrophobicity distribution profiles for 1CMA chain A. Circles mark residues involved in interaction with other compounds in the complex: magenta – chain B (P–P), green – ligand, yellow – DNA. Red background indicates a discordant fragment (88–99), which when eliminated the RD value decreases below 0.5.

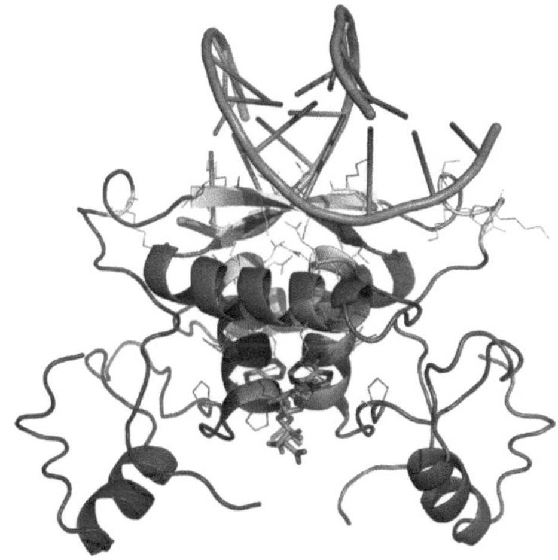

Fig. 6.B.4 3D presentation of 1CMA: chain (A) blue, chain (B) gray. Residues engaged in protein-ligand (green) and protein-nucleic (yellow) interactions have side chains displayed. The fragments marked in each chain in red (88–99) are locally discordant.

When considering the dimer as a whole, each chain is also regarded as accordant, which shows that both chains cooperate to build a shared hydrophobic core. Again, the information required in the process of complexation is expressed as discrepancies between T and O in each individual chain. Note that dimerization is not the ultimate goal of the complex, and that it also includes residues which mediate its biological activity – i.e. DNA binding.

The above conclusions remain valid both for the apo form of the presented protein and for its complex with the co-repressor, suggesting that the FOD status is related to the biological role of the complex. A discussion of various types of relations between the monomer and the dimer in light of the fuzzy oil drop model can be found in Ref. [4].

The status of the homodimer suggests that this structure emerges as a result of interactions with the water environment, which produces a shared hydrophobic core. When analyzed as part of the dimer, chain A appears to contribute to the shared core, while chain B is discordant on its own and does not exhibit accordance when viewed as part of the complex. Balanced values of correlation coefficients (TvO and HvO) indicate a synergistically optimized structure. It is also interesting to note the very low RD value computed for the interface, showing that the interface area is a major stabilizing factor, enabling the dimer to retain a shared core. Eliminating interface residues from FOD calculations produces a much more discordant dimer structure, with a notable increase in the value of RD. It follows that the dimer functions as a coherent structural unit, stabilized by its own hydrophobic core to which both chains contribute.

Individual chains

Each chain, when analyzed on its own, appears to diverge from the monocentric distribution of hydrophobicity. This is linked to exposure of strongly hydrophobic residues on the surface, contrary to the FOD model. Notably, the exposed residues belong to the complexation interface and are expected to encode information required in the complexation process. This conclusion is supported by calculation of RD values for interface fragments: eliminating such residues from computations results in a significant decrease in RD, although the remainder of each chain is still regarded as discordant.

The activity profile of the presented protein involves DNA binding. Residues which mediate this process conform to the model since they exhibit high polarity and are exposed on the surface, in agreement with the micellar protein structure model. Eliminating these residues results in a higher value of RD for chain A (when treated as a separate unit).

DNA binding

Identification of residues which cause RD to exceed 0.5 despite prior elimination of discordant interface residues may yield clues regarding the protein's capability for structural rearrangement. In the scope of the individual chain, discordance is caused by the fragments at 45–47, 68–72, 79–83 and 95–100 (in addition to the aforementioned interface fragments). All these fragments belong to outlying loops and may potentially undergo conformational changes during DNA complexation.

The sketch included in the chapter title illustrates the specific conditions discussed above. A detailed analysis of discrepancies between the status of individual monomers and the resulting dimers (in all possible variants) is presented in Refs. [4–7].

References

[1] Somers WS, Phillips SEV. Crystal structure of the met represser–operator complex at 2.8 Å resolution reveals DNA recognition by β-strands. Nature 1992;359(6394): 387–93. https://doi.org/10.1038/359387a0.

[2] Rafferty JB, Somers WS, Saint-Girons I, Phillips SEV. Three-dimensional crystal structures of *Escherichia coli* met repressor with and without corepressor. Nature 1989; 341(6244):705–10. https://doi.org/10.1038/341705a0.

[3] Janin J, Henrick K, Moult J, Eyck LT, Sternberg MJE, Vajda S, Wodak SJ. CAPRI: a critical assessment of PRedicted interactions. Proteins: Structure, Function, and Genetics 2003;52(1):2–9. https://doi.org/10.1002/prot.10381.
[4] Dygut J, Kalinowska B, Banach M, Piwowar M, Konieczny L, Roterman I. Structural interface forms and their involvement in stabilization of multidomain proteins or protein complexes. International Journal of Molecular Sciences 2016;17(10):1741. https://doi.org/10.3390/ijms17101741.
[5] Marchewka D, Jurkowski W, Banach M, Roterman-Konieczna I. Prediction of protein-protein binding interfaces. Focus on Structural Biology 2012:105–33. https://doi.org/10.1007/978-94-007-5285-6_6.
[6] Banach M, Konieczny L, Roterman-Konieczna I. Can the structure of the hydrophobic core determine the complexation site? Focus on Structural Biology 2012:41–54. https://doi.org/10.1007/978-94-007-5285-6_3.
[7] Banach M, Konieczny L, Roterman-Konieczna I. Use of the "fuzzy oil drop" model to identify the complexation area in protein homodimers. Protein Folding in Silico 2012: 95–122. https://doi.org/10.1533/9781908818256.95.

CHAPTER 6.C

Ligand binding cavity encoded as a local hydrophobicity deficiency

Mateusz Banach[1], Leszek Konieczny[2], Irena Roterman[1]
[1]Department of Bioinformatics and Telemedicine, Jagiellonian University—Medical College, Krakow, Poland
[2]Chair of Medical Biochemistry, Jagiellonian University—Medical College, Krakow, Poland

Contents

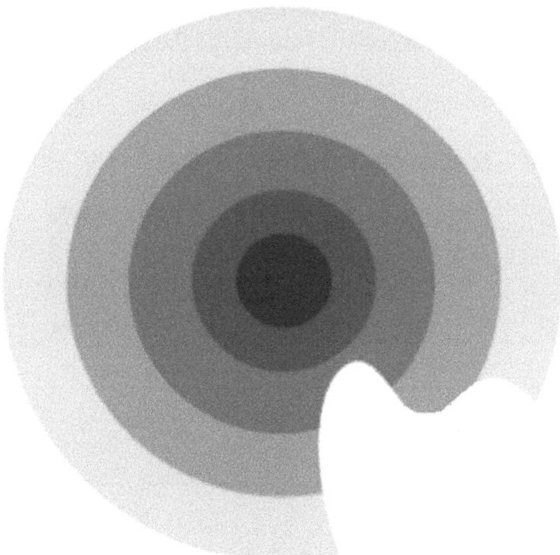

The above diagram illustrates the specific deviation from the theoretical distribution of hydrophobicity which is associated with the presence of a ligand binding cavity.

An example of strong affinity of a polypeptide chain for a specific "alien" (non-protein) molecule is provided by erythrocruorin (PDB ID: 1ECD) [1]. We note the characteristic presence of a strongly hydrophobic ligand – heme,

From Globular Proteins to Amyloids
ISBN: 978-0-08-102981-7
https://doi.org/10.1016/B978-0-08-102981-7.00009-9

located centrally in a β-barrel structure (CATH code 1.10.490.10 – mainly alpha orthogonal bundle).

Erythrocruorin (1ECD) is a large oxygen-carrying protein. PDB lists its monomeric structure in complex with heme. Its FOD status is presented in Table 6.C.1. As shown, the protein as a whole does not conform to the theoretical distribution of hydrophobicity. Eliminating residues which participate in ligand binding lowers the overall value of RD; however, the value remains greater than 0.5.

Ligand binding residues are characterized by particularly high values of RD, indicating strong deviation from the monocentric Gaussian distribution. Further analysis of theoretical and observed profiles reveals other discordant residues, which do not participate in protein-ligand interactions. Eliminating such residues from calculations produces an RD value lower than 0.5 for the remainder of the chain. This enables us to identify fragments which adopt a micellar conformation and stabilize the protein.

Table 6.C.1 Fuzzy oil drop parameters for 1ECD. The status is presented for the entire chain, for ligand-binding residues (Lig), for parts of the chain following elimination of ligand-binding residues (No Lig) and for the remainder of the chain following elimination of other discordant fragments (i.e. for fragments which ensure structural stabilization).

		RD		Correlation coefficient		
Erythrocruorin (1ECD)	**Fragment**	**T-O-R**	**T-O-H**	**HvT**	**TvO**	**HvO**
		0.560	0.377	0.509	0.520	0.760
Lig		**0.629**	0.391	0.653	0.205	0.656
No Lig		**0.522**	0.350	0.459	0.569	0.676
Residues eliminated	51–54 114–117	0.468	0.298	0.491	0.642	0.803

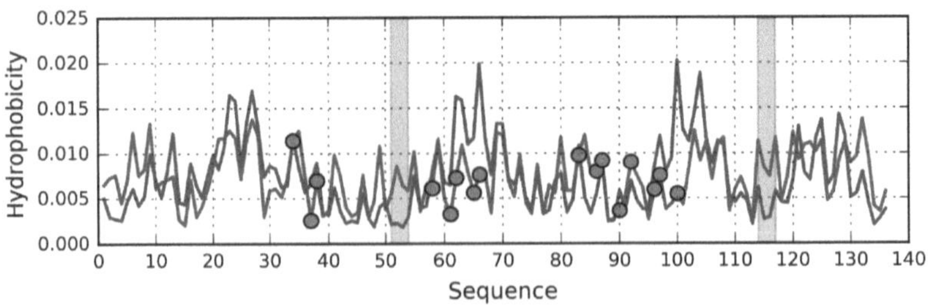

Fig. 6.C.1 Theoretical (T, blue) and observed (O, red) hydrophobicity distribution profiles for 1ECD. Green circles mark residues involved in P-P interaction. Red background indicates a discordant fragment (51–54, 114–117).

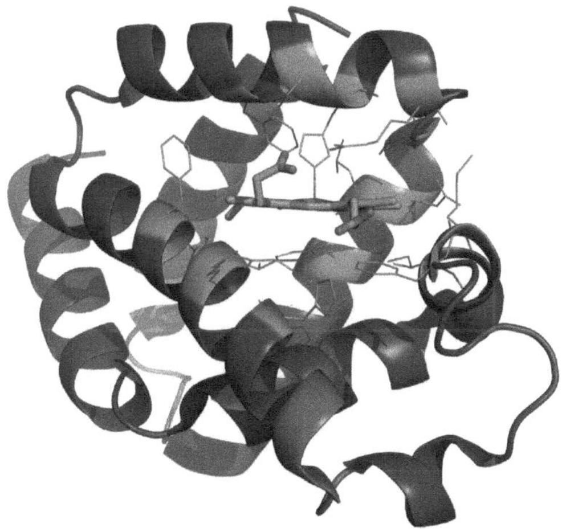

Fig. 6.C.2 3D presentation of 1ECD. Residues engaged in interactions between protein and heme are highlighted in green and have side chains displayed. Fragments of exposed hydrophobicity (51−54, 114−117) are shown in red.

It should be noted that there is a notable exposure of hydrophobicity at positions 51−54, possibly indicating a complexation interface. The biologically active form of erythrocruorin, identified *in vivo*, is a complex consisting of multiple chains.

Correlation coefficients appear balanced, suggesting that intrinsic hydrophobicity is not a dominant factor, and that the molecule generally adopts a micellar form (Figs. 6.C.1 and 6.C.2).

Further information concerning the structure of ligand-binding cavities can be found in Refs. [2−4].

References

[1] Steigemann W, Weber E. Structure of erythrocruorin in different ligand states refined at 1·4 Å resolution. Journal of Molecular Biology 1979;127(3):309−38. https://doi.org/10.1016/0022-2836(79)90332-2.

[2] Kalinowska B, Banach M, Wiśniowski Z, Konieczny L, Roterman I. Is the hydrophobic core a universal structural element in proteins? Journal of Molecular Modeling 2017; 23(7). https://doi.org/10.1007/s00894-017-3367-z.

[3] Banach M, Konieczny L, Roterman-Konieczna I. Ligand-binding-site recognition. Protein Folding in Silico 2012:79−93. https://doi.org/10.1533/9781908818256.79.

[4] Alejster P, Banach M, Jurkowski W, Marchewka D, Roterman-Konieczna I. Comparative analysis of techniques oriented on the recognition of ligand binding area in proteins. Focus on Structural Biology 2012:55−86. https://doi.org/10.1007/978-94-007-5285-6_4.

CHAPTER 7

Solenoid — An amyloid under control

Mateusz Banach[1], Irena Roterman[1]
[1]Department of Bioinformatics and Telemedicine, Jagiellonian University—Medical College, Krakow, Poland

Contents

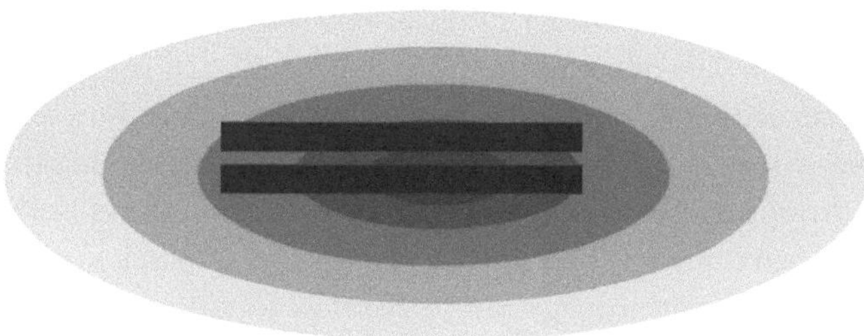

The diagram illustrates a structure which is consistent with the fuzzy oil drop model despite containing a solenoid fragment, characterized by linear propagation of alternating bands of variable hydrophobicity. This unusual structure, when "packaged" in a soluble envelope, is protected against unrestricted aggregation.

The solenoid is a supersecondary structure shaped like a tubule, with three or four "walls" consisting of short β-strands generating elongated parallel one to each other β-sheets. This unusual structure is sometimes also called a β-helix. The symbol "β" refers to the presence of β-strands in each segment, while "helix" describes the general characteristics of the

From Globular Proteins to Amyloids
ISBN: 978-0-08-102981-7
https://doi.org/10.1016/B978-0-08-102981-7.00010-5

solenoid as a whole, since solenoids are usually slightly twisted about their main axis.

An α-helix is stabilized by a network of hydrogen bonds which run parallel to its axis. These bonds join every third or fourth amino acid. In a solenoid, numerous hydrogen bonds are found within the system of β-strands forming β-sheets which make up the walls of the structure. Each of these bonds links residues which are much further apart in the chain than in the case of an α-helix (depending on the overall breadth of the solenoid and on the optional presence of exposed loops between each unit segment).

Entering "solenoid" in the PDB search interface produces two groups of proteins: antifreeze proteins and lyases. In the former group, solenoid fragments tend to be regularly structured, while in lyases they are often accompanied by numerous additional elements such as helical folds, loops (often quite long), random coils and even domain-like fragments. Thus, we will illustrate our discussion of solenoids with two sample proteins representing both groups.

The structural peculiarity of solenoids immediately raises the question – how can such a structure emerge in the first place? It turns out that each solenoid contains bands of bands of variable hydrophobicity which propagate along its main axis. The arrangement of bands appears to promote formation of long tubular structures, consisting of three or four β-sheets (four or sometimes three) with a more or less symmetrical cross section.

In a helix the radius of curvature is far smaller than in β-solenoids. Additionally, helices are more regular than solenoid fragments. Based on specific examples we will attempt to show that alignment of bands characterized by variable hydrophobicity is the main driving force behind formation of elongated structural forms. The linear propagation of bands is characterized by the possible no-limit propagation in contrast to globular 3 D Gauss-like structural forms. This is why solenoid as any other structure based on linear propagation of any character requires specific "caps" to stop unlimited propagation. This aspect will be discussed using the examples of proteins with solenoids in their structures.

Antifreeze proteins represented by protein extracted from *Flavobacterum frigoris PS1* found in the Antarctic region

As an example of the antifreeze family, we will consider a protein extracted from *Flavobacterum frigoris PS1* found in the Antarctic region

(PDB ID: 4NU2) [1]. This protein was selected due to its strong topological resemblance to the lyase which will be presented later on.

The antifreeze protein (4NU2) consists of a solenoid fragment along with several outlying fragments forming a package: a short "stop" helix (also referred to as "cap") attached to a terminal section of the solenoid and preventing unrestricted complexation, another helix which runs parallel to the solenoid. The protein consists also several random coil fragments.

The molecule as a whole does not conform to the Gaussian distribution of hydrophobicity (Table 7.1). Likewise, the solenoid itself (which in this case consists of three β-sheets) is strongly discordant. The sheet identified as

Table 7.1 FOD parameters (RD values and correlation coefficients) for antifreeze protein (4NU2). Results are listed for the complete molecule, for the β-sheets comprising the solenoid, for the solenoid as a whole and for other fragments of the chain, including selected secondary folds. Non-solenoid – parts of the chain which do not comprise the solenoid; 70% of non-solenoid – part of the chain which remain accordant with the expected distribution of hydrophobicity (detailed description given in the text). Values listed in boldface indicate discordance.

Antifreeze protein (4NU2)	Fragment	RD		Correlation coefficient		
		T-O-R	T–H	HvT	TvO	HvO
Chain	61–276	0.669	0.517	0.290	0.470	0.642
Solenoid						
β-sheet	**73–77***	**0.744**	**0.518**	**0.113**	**0.402**	**0.483**
β-sheet	81–84*	0.338	0.392	0.538	0.802	0.816
β-sheet	88–90*	0.718	0.550	0.181	0.169	0.805
β-sheet	166–167*	0.725	0.667	0.829	0.457	0.623
β-sheet	**88–90 + 166–167**	**0.808**	**0.694**	**0.163**	**0.157**	**0.745**
Complete		**0.660**	0.437	0.302	0.490	0.675
Other fragments						
Stop – random coil	104–112	0.298	0.110	0.426	0.761	0.828
Stop – helix	153–164	0.337	0.197	0.478	0.741	0.271
Non-solenoid		**0.616**	0.432	0.246	0.370	0.644
70% non-solenoid		0.470	0.320	0.352	0.632	0.658
Parallel helix	128–148	0.479	0.477	0.452	0.632	0.697
RC (parallel)	61–73	0.292	0.375	0.389	0.810	0.601
SS-bond	107–124	**0.672**	0.494	0.233	0.466	0.655

* only numbers of residues belonging starting (first) Beta-strand in the Beta sheet are given

73—77 (values corresponding to the location of its initial β-strand) is particularly amyloid-like, with high RD and dominant HvO correlation coefficient.

The solenoid is defined in this work as set of β-strands forming the helical form. The β-strands together with one additional residue at each end of strand is taken to define the solenoid part of the protein. This notation will be used all over this work.

A single disulfide bond (107C—124C) is present and appears to structurally reinforce the protein.

As already discussed, the solenoid is a linear structure. Producing such a structure calls for a mechanism which promotes elongation of a tubular form. According to observations involving other antifreeze proteins which include solenoid fragments, we assume that this mechanism is provided by a specific distribution of hydrophobicity, deviating from the monocentric Gaussian in favor of a linear pattern where bands of high and low hydrophobicity are arranged in an alternating fashion.

The profiles shown in Fig. 7.1A make it easy to identify the solenoid, which is characterized by a sinusoidal pattern of hydrophobicity, along with strong deviations from T. The C-terminal fragment, where

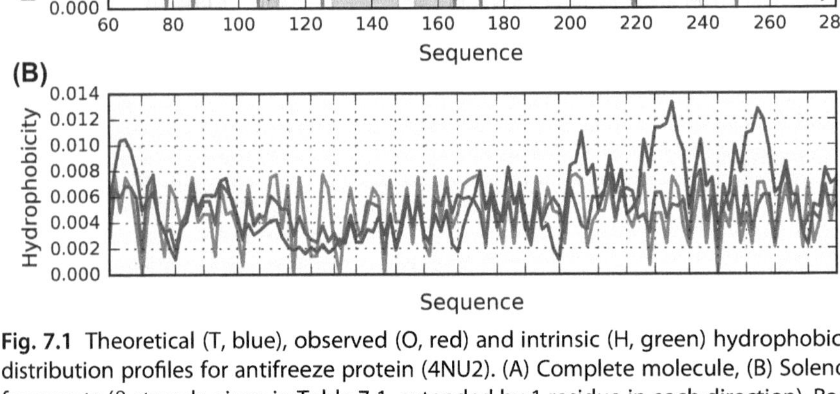

Fig. 7.1 Theoretical (T, blue), observed (O, red) and intrinsic (H, green) hydrophobicity distribution profiles for antifreeze protein (4NU2). (A) Complete molecule, (B) Solenoid fragments (β-strands given in Table 7 1, extended by 1 residue in each direction). Background colors highlight the position of solenoid (yellow), parallel helix (red) and "stop" fragments (green). To avoid displaying multiple gaps between the fragments of the solenoid, profiles in B are conflated into contiguous sequence. However ticks on the horizontal axis can be used to determine the location of those fragments within the complete sequence.

hydrophobicity is expected to remain strong, does not differ significantly from the remainder of the chain and is therefore discordant. Plotting the distribution of hydrophobicity for the solenoid itself (Fig. 7.1B) reveals a characteristic sinusoidal pattern composed of alternating bands. Clearly, this pattern does not correspond to the theoretical monocentric distribution of hydrophobicity. Another characteristic property of solenoids is the strong convergence between O and H (Fig. 7.1B, further confirmed by the data in Table 7.1). This strong correlation HvO with low correlation for HvT and especially TvO suggests the dominating role of intrinsic hydrophobicity of each residue. It is in contrast to expected uni-centric hydrophobic core formation which requires the synergy of all residues.

Taken together, it will shown later on that the structural properties of solenoid fragments resemble those which characterize amyloid fibrils.

This linear propagation — especially in the context of β-sheets — is capable of unrestricted growth, given that both edges of the solenoid fragment are able to form H-bond systems. Consequently, a dedicated "stop" signal is required to prevent the solenoid from growing without bound. Following analysis of a large set of solenoid-containing proteins we find that the signal assumes the form of a short helical "cap", which is typically highly consistent with the fuzzy oil drop model (low RD values). This means that the cap must be amphipathic: on its underside it remains compatible with the outermost segment of the solenoid, while on the outside it ensures suitable conditions for interaction with the aqueous solvent. Once in place, it effectively arrests further linear propagation. In later sections we will again refer to capping helices as a potential targets in the search for new drugs capable of preventing growth of amyloid structures. It will be shown in the Chapter 10.

Another characteristic property of solenoid-containing proteins is the presence of a long helix which runs parallel to the solenoid fragment and remains consistent with the Gaussian distribution. This helix also mediates entropically advantageous contact with the aqueous environment: it promotes solubility by preventing face-on complexation (which might otherwise be possible given the solenoid's discordant status). The helix is typically closely aligned with a discordant β-sheet — in our case with the sheet identified as 72—77.

Other components of the "packaging" (which refers to fragments other than the solenoid itself) exhibit variable FOD status. Altogether, 70% of the non-solenoid part of the chain remains consistent with the model and appear to promote solubility. We identify this part by progressively eliminating discordant residues until we obtain RD < 0.5. Some additional discordances may be explained by the dynamic stereochemistry of outlying loops, which may locally deviate from the Gaussian distribution. According to the fuzzy oil drop model, antifreeze proteins perform their function by enforcing a nonstandard structural arrangement of nearby water molecules, disfavoring the formation of ice crystals. This disruption — while not yet precisely defined — may explain the need for a solenoid fragment, with its peculiar distribution of hydrophobicity. Experimental research suggests that water effectively "levitates" on top of hydrophobic surfaces [2], while the structure of the solvent adjacent to hydrophilic bands is likely determined by interactions between H_2O molecules and polar groups. Interestingly, the mobility of water molecules is observed to increase in closeness, proximity to antifreeze proteins [3]. A broader selection of antifreeze proteins is analyzed in Ref. [4].

An alternating pattern of variably hydrophobic bands appears optimal given the expected role of antifreeze proteins. In the Chapter 5. Which discuss the properties of FOD-compliant molecules we mention, among others, type III antifreeze proteins. Their surface is entirely composed of polar groups, which also affects the surrounding solvent and prevents crystallization. A similar mechanism may be proposed for other small organic molecules such as saccharides, lipids or even salt, which is a popular de-icing agent. Fig. 7.2 illustrates the location of various fragments listed in Table 7.1.

Lyase family represented by pectate lyase

The lyase family is represented by pectin and pectate lyase E.C. 4.2.2.2 isolated from *Bacillus subtilis* [5]. This enzyme catalyzes eliminative cleavage

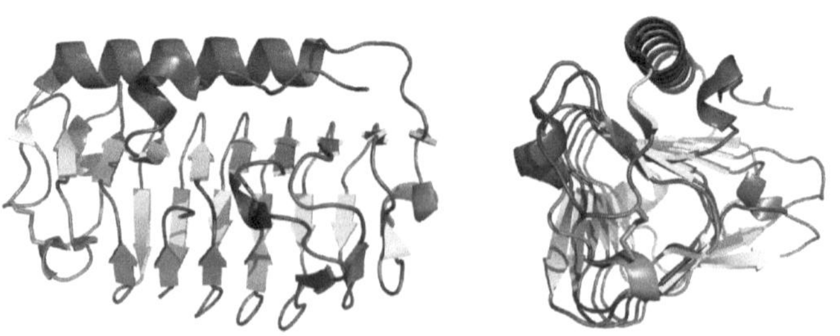

Fig. 7.2 3D presentation of antifreeze protein (4NU2) in two perspectives. Colors distinguish different fragments of the molecule: solenoid — yellow, parallel helix — red, "stop" fragments — green, disulfide — orange.

of pectate to give oligosaccharides with 4-deoxy-alpha-D-gluc-4-enuronosyl groups at their non-reducing ends. The structure is listed in PDB under ID 1BN8.

The presence of a Ca^{2+} ion close to an arginine residue appears important for catalysis since it is conserved across all pectin and pectate lyases. The protein consists of 399 residues and contains no disulfide bonds. Catalytic residues are identified as 184D, 206R and 279R.

Lyase (1BN8) is a very complex enzyme, both from the point of view of topology and intended function. Consequently, our analysis must address several distinct issues: its enzymatic activity and specific conformational properties, which are characterized in Table 7.2.

The molecule as a whole appears to lack a central hydrophobic core. Comparison of correlation coefficients indicates strong involvement of intrinsic hydrophobicity in shaping the tertiary conformation of lyase (1BN8) (Table 7.2 and Fig. 7.3).

Eliminating catalytic residues produces a slight decrease in RD, suggesting that these residues deviate from the theoretical distribution (although not very strongly). Further analysis reveals that discordances are centered

Table 7.2 FOD parameters (RD values and correlation coefficients) for lyase (1BN8). Results are listed for the complete molecule, for the β-sheets comprising the solenoid, for the solenoid as a whole and for other fragments of the chain, including selected secondary folds. Values listed in boldface indicate discordance. Amyloid-like fragments: bold — strong discordance between T and O including these with biased relation of correlation coefficients — the relation HvO dominating.

		RD		Correlation coefficient		
Lyase (1BN8)	**Fragment**	**T-O-R**	**T-O-H**	**HvT**	**TvO**	**HvO**
Chain	1–399	**0.684**	**0.559**	0.194	0.346	0.749
No E res.	184D, 206R, 279R	**0.680**	**0.553**	0.211	0.361	0.749
184D	±5 aa	**0.529**	**0.553**	−0.064	0.333	0.775
206R	±5 aa	0.482	**0.654**	0.360	0.711	0.880
279R	±5 aa	**0.505**	0.353	0.212	0.231	0.865
All cat. res.	**162–280**	**0.563**	0.425	0.188	0.418	0.730
Solenoid						
β-sheet	32–35*	**0.696**	0.361	0.350	0.414	0.709
β-sheet	61–63*	**0.705**	**0.538**	0.132	0.079	0.892
β-sheet	**122–125***	**0.730**	**0.628**	**−0.135**	**−0.177**	**0.655**
Complete		**0.727**	**0.529**	**0.121**	**0.148**	**0.773**

(*Continued*)

Table 7.2 FOD parameters (RD values and correlation coefficients) for lyase (1BN8). Results are listed for the complete molecule, for the β-sheets comprising the solenoid, for the solenoid as a whole and for other fragments of the chain, including selected secondary folds. Values listed in boldface indicate discordance. Amyloid-like fragments: bold – strong discordance between T and O including these with biased relation of correlation coefficients – the relation HvO dominating.—cont'd

		RD		Correlation coefficient		
Lyase (1BN8)	**Fragment**	**T-O-R**	**T-O-H**	**HvT**	**TvO**	**HvO**
Other fragments						
β-sheet	168-170*	**0.599**	**0.392**	**−0.161**	**−0.073**	**0.873**
Random coil	1–31	0.469	0.417	0.310	0.642	0.564
Stop – helix	38–46	0.323	0.642	0.773	0.719	0.922
Loop*	63–120	0.640	0.462	0.063	0.252	0.847
Loop*	162–184	0.763	0.682	-0.256	-0.161	0.711
Loop*	203–224	0.684	0.486	-0.093	0.343	0.751
Loop*	297–309	0.690	0.707	0.533	0.739	0.642
RC	327–331	0.035	0.088	0.790	0.993	0.853
RC	354–364	0.416	0.396	0.100	0.657	0.469
Stop – helix	358–364	0.603	0.562	-0.182	0.447	0.548
RC	365–381	0.673	0.748	0.215	0.113	0.761
Helix – parallel	382–393	0.498	0.283	0.390	0.573	0.852
RC	393–399	0.189	0.062	0.433	0.940	0.557
Loops*		0.735	0.610	0.080	0.253	0.802
Helix	2–6	0.510	0.396	0.418	0.352	0.726
	13–17	0.304	0.142	−0.379	0.866	−0.648
	74–79	0.253	0.100	0.801	0.812	0.917
	84–92	0.250	0.184	0.873	0.809	0.922
	93–98	0.675	0.441	−0.198	0.282	0.869
	104–121	0.723	0.551	0.072	0.163	0.846
	207–211	**0.866**	**0.789**	**−0.983**	**−0.742**	**0.835**
	257–261	0.715	0.095	−0.680	0.310	0.276
	385–393	0.352	0.179	0.437	0.679	0.896

* only numbers of residues belonging starting (first) Beta-strand in the Beta sheet are given

upon residues Asp184 and Arg279, along with their immediate neighborhood (5 aa in each direction). Notably, the neighborhood of Asp184 is characterized by a negative value of HvT, indicating that this fragment actively opposes the formation of a hydrophobic core. The larger fragment which contains all catalytic residues (162–280) is likewise discordant and reveals strong involvement of intrinsic hydrophobicity. Given the fragment's role in facilitating catalysis, its status should be regarded as important.

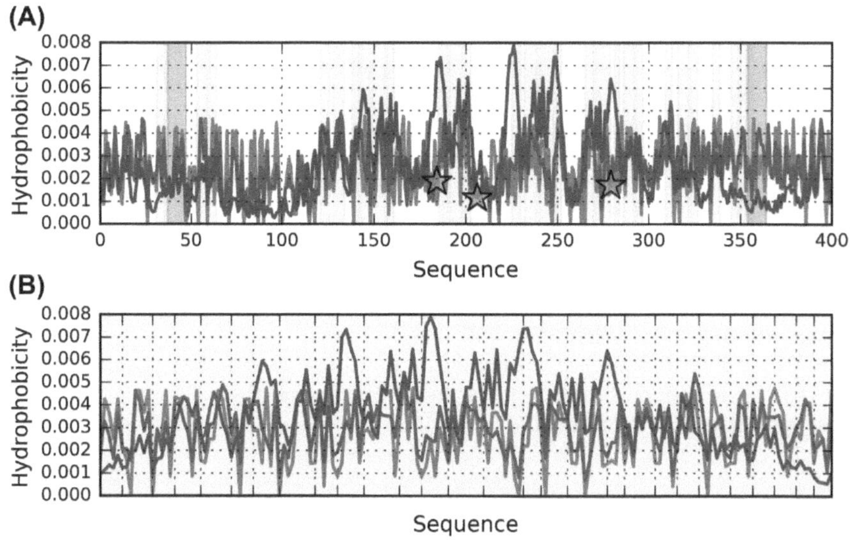

Fig. 7.3 Theoretical (T, blue), observed (O, red) and intrinsic (H, green) hydrophobicity distribution profiles for lyase (1BN8). (A) Complete molecule, (B) Solenoid fragments (β-strands given in Table 7.2, extended by 1 residue in each direction). Background colors highlight the position of solenoid (yellow) and "stop" fragments (green). Orange stars mark positions of catalytic residues. To avoid displaying multiple gaps between the fragments of the solenoid, profiles in B are conflated into contiguous sequence. However ticks on the horizontal axis can be used to determine the location of those fragments withing the complete sequence.

As previously postulated in the Chapter 3, which discusses encoding of information required to ensure biological activity, a highly specific catalytic active site must contain a large quantity of information. This explains the relatively strong discordance between the observed distribution of hydrophobicity and the corresponding deterministic (micelle-like) distribution.

When analyzing the solenoid as a whole, we note that it diverges from the theoretical Gaussian in favor of a distribution dominated by the intrinsic properties of its component residues. In this scope, the β-sheet starting at 122—125 is strongly amyloid-like, with high values of RD along with negative values of both HvT and TvO. Its structure not only diverges from the Gaussian distribution, but actively opposes it.

Fig. 7.4 illustrates the profiles of each fold comprising the 122—125 β-sheet (which is itself part of the solenoid fragment).

The values of FOD parameters describing this β-sheet are consistent with the situation presented in Fig. 7.4A steady decrease in hydrophobicity is predicted by the fuzzy oil drop model (since each successive fold is closer to the

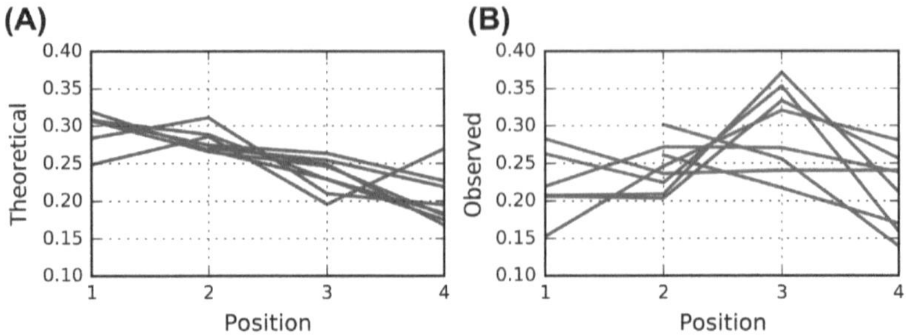

Fig. 7.4 Theoretical (T, blue) and observed (O, red) hydrophobicity distribution profiles for lyase (1BN8) β-sheet which starts with 122–125 strand. (A) Theoretical hydrophobicity, (B) Observed hydrophobicity. Numbers on horizontal axes denote positions of residues relative to sequence fragments of the β-strands after their alignment with their spatial distribution in the β-sheet.

surface of the protein), whereas the observed distribution is characterized by two local maxima separated by a distinct local minimum. The rightmost maximum clearly opposes the theoretical distribution. Thus, Fig. 7.4 reveals the presence of two strongly hydrophobic bands separated by a hydrophilic band. The fact that this band-like pattern is replicated in successive folds, despite differences in their composition, is the hallmark of an amyloid-like structure (this will be further discussed in the next chapter).

It seems that alternating bands of variable hydrophobicity are the driving force behind self-assembly of fibrillar structures. Such structures are, in principle, capable of unrestricted growth, for example by forming complexes with additional lyases. To prevent this undesirable phenomenon, the fragment at 38–46 acts as a "stopper" by presenting a steric obstacle: the helix effectively "caps" the solenoid and enables contact with water thanks to its amphipathic character. The opposite helix at 358–364 lacks these properties, which likely means that it plays a different role, and that a single "cap" is sufficient to prevent unlimited elongation of the solenoid. A similar situation is observed in the previously discussed antifreeze protein.

The band-like pattern exhibited by the 122–125 β-sheet may induce other important effects: for example, it is plausible that it enforces a specific local ordering of water molecules in order to promote catalysis. Similarly, the presence of a Ca^{2+} ion close to the active site may serve to create a suitable force field in terms of electrostatic interactions (the ion, by itself, does not disrupt the distribution of hydrophobicity in its neighborhood). Other fragments – especially random coils – appear to mediate interaction with the aqueous solvent, given that most of them are found on the surface of the protein and effectively "package" the solenoid.

The status of helical fragments varies, with notable discordances observed in helices which surround the active site (93—98, 104—121, 207—211). Assuming that the molecule as a whole must be able to generate an internal force field which facilitates its biological activity, and that the surrounding solvent must also play a role in this process, these three helices may be regarded as crucially important. Even though our presentation focuses on a single representative lyase, analysis of a larger set of solenoid-containing lyases supports our conclusions. In particular, the properties of the internal force field along with the postulated involvement of the enzyme's aqueous environment (which is acted upon by the protein itself), appear to be of fundamental importance for its biological activity.

Fig. 7.5 visualizes all key fragments of the presented lyase.

In conclusion, it is interesting to note the linear structure of the solenoid fragment, which closely resembles conditions found in amyloids. Considering the threat of unlimited elongation, biologically active solenoids must be equipped with appropriate "caps". The role and structure of such fragments will be further discussed in the Chapter 11. Which presents the potential development of drugs capable of preventing unrestricted growth of amyloid fibrils.

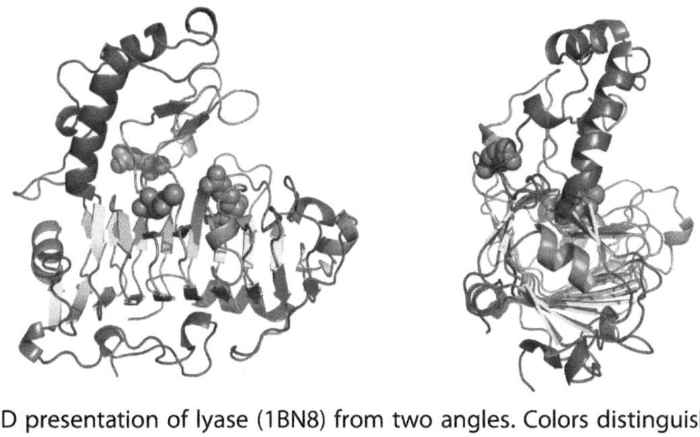

Fig. 7.5 3D presentation of lyase (1BN8) from two angles. Colors distinguish different fragments of the molecule: solenoid — yellow, "stop" fragments — green, neighborhood of catalytic site — salmon (also including parts of the solenoid), orange spheres — catalytic residues, teal sphere — Ca^{2+} ion.

"Perfect" solenoid

A "perfect" solenoid structure is embodied by the antifreeze protein from *Tenebrio monitor*, a beetle species (PDB ID: 1EZG) [6]. The protein consists of 81 residues, including a tandem of 12-residue repeats (TCT_S__C__A_). The regularity of its SS bond structure is striking. FOD analysis reveals that the protein deviates from T in favor of H (RD > 0.5) and that therefore its tertiary conformation is determined by the intrinsic hydrophobicity of its residues.

A perfectly symmetrical network of SS bonds (8 in total) can found in the central part of the protein. Consequently, we may speak of a "quasi" hydrophobic core which spans the entire length of the solenoid. The protein also includes several exposed loops consisting of polar residues and likely mediating contact with the solvent (in a manner consistent with the fuzzy oil drop model). Nevertheless, plotting T and O profiles for the entire chain reveals a prominent sinusoidal pattern of alternating maxima and minima, stretching from the N-terminal fragment all the way to the C-terminal fragment.

The distributions plotted in Fig. 7.6 reveal consistent placement of local maxima which appear accordant with the T distribution. The C-terminal fragment is exception representing significant discordance in respect to T distribution.

Another striking property is the near-perfect accordance of fragments bracketed by disulfide bonds. Much like the hydrophobic core, disulfides are thought to contribute to tertiary structural stabilization of proteins. It seems that in the case of 1EZG both factors work in tandem (Figs. 7.7 and 7.8).

The presented protein, which exhibits linear propagation of alternating bands of high and low hydrophobicity, is a good example of how evolution has devised ways to prevent uncontrolled growth of fibrillar structures (which would otherwise be possible in the absence of the N- and C-terminal fragments). The authors of [6] remark that this repetitive sequence translates into an exceptionally regular β-helix, perhaps the most regular protein structure yet observed. Despite the lack of a centric core, the solenoid exhibits strong alignment between the observed and theoretical hydrophobicity, with only its N- and C-terminal fragments remaining locally discordant (Fig. 7.6). The concentration of Cys residues (high intrinsic hydrophobicity) in the central part of the solenoid produces an elongated *quasi*-core, which explains the good agreement between T and O.

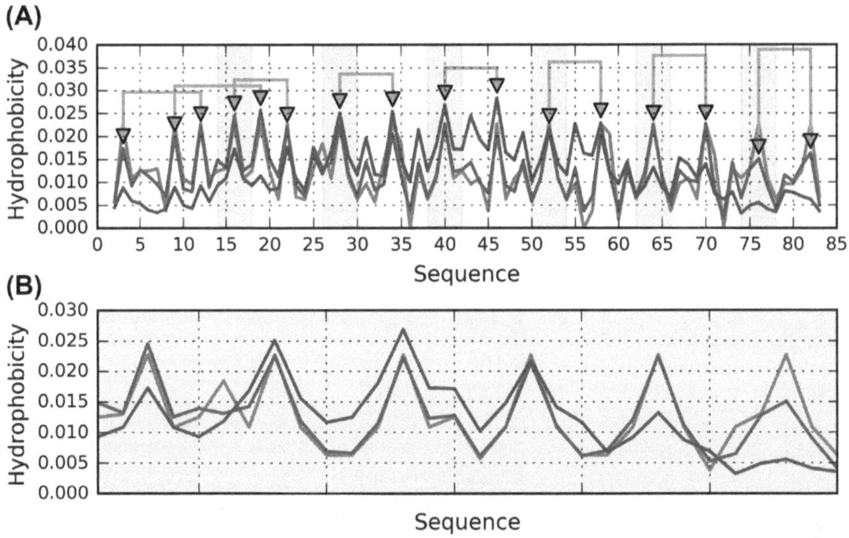

Fig. 7.6 Theoretical (T, blue), observed (O, red) and intrinsic (H, green) hydrophobicity distribution profiles for antifreeze protein (1EZG). (A) Complete molecule, (B) Solenoid fragments (β-strands given in Table 7.3, extended by 1 residue in each direction). Background colors highlight the position of solenoid (yellow). Orange "arrows" correspond to SS bonds formed by Cys residues. To avoid displaying multiple gaps between the fragments of the solenoid, profiles in B are conflated into contiguous sequence. However ticks on the horizontal axis can be used to determine the location of those fragments withing the complete sequence.

From the point of view of the fuzzy oil drop model, this protein is unique in terms of the correlation coefficients computed for successive β-strands and fragments bracketed by disulfide bonds. In all three cases, values of the correlation coefficients are greater than 0.9, indicating near-perfect alignment between the sequence and the structure of the protein. Good agreement between T and O is thought to result from excellent alignment between H and both T and O. Consequently, the protein may be viewed as an example of structural properties encoded in the amino acid sequence. Its antifreeze properties emerge as a result of mechanisms similar to those discussed in Chapter 5: the surface is covered in polar groups which alter the structural properties of the surrounding solvent and disfavor formation of ice crystals. Likewise, the solenoid present in other antifreeze protein (4NU2) exposes alternating bands of high and low hydrophobicity which disrupt the natural structure of the solvent, preventing crystallization.

Table 7.3 FOD parameters (RD values and correlation coefficients) for antifreeze protein (1EZG). Results are listed for the complete molecule, for the β-sheet comprising the solenoid and for fragments bracketed by disulfide bonds. Values listed in boldface indicate discordance.

		RD		Correlation coefficient		
Antifreeze protein (1EZG)	**Fragment**	**T-O-R**	**T-O-H**	**HvT**	**TvO**	**HvO**
Chain	**2–83**	**0.559**	**0.652**	**0.314**	**0.485**	**0.885**
Solenoid						
β-strand	14–18	0.156	0.831	0.945	0.922	0.987
β-strand	26–30	0.183	0.422	0.646	0.916	0.874
β-strand	38–42	0.118	0.879	0.985	0.989	0.995
β-strand	50–54	0.168	0.957	0.994	0.991	0.998
β-strand	62–66	0.202	0.871	0.989	0.994	0.998
β-strand	74–78	0.259	0.650	0.850	0.953	0.895
β-sheet		0.473	0.850	0.508	0.658	0.922
Complete		0.543	0.838	0.332	0.540	0.890
SS-bonds						
	3–12	0.187	0.518	0.866	0.902	0.943
	9–19	0.287	0.788	0.601	0.751	0.950
	16–22	0.338	0.703	0.804	0.723	0.973
	28–34	0.070	0.188	0.831	0.979	0.902
	40–46	0.237	0.416	0.797	0.879	0.950
	52–58	0.344	0.275	0.662	0.779	0.980
	64–70	0.366	0.803	0.546	0.597	0.980
	76–82	0.411	0.624	0.077	0.514	0.855

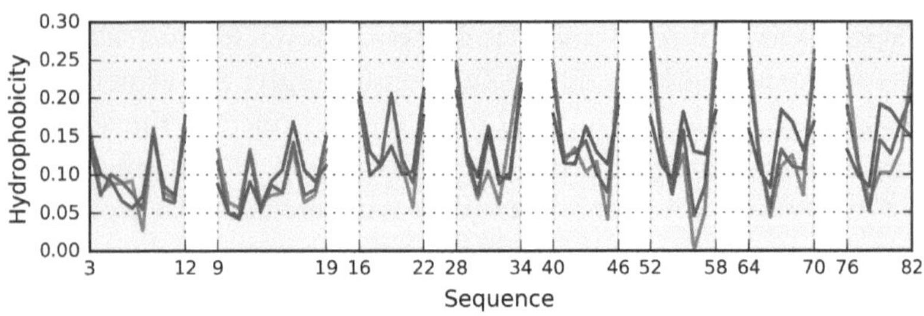

Fig. 7.7 Theoretical (T, blue), observed (O, red) and intrinsic (H, green) hydrophobicity distribution profiles for antifreeze protein (1EZG) fragments bracketed by disulfide bonds. Note that these fragments overlap in the complete sequence, as denoted by Cys positions on horizontal axis.

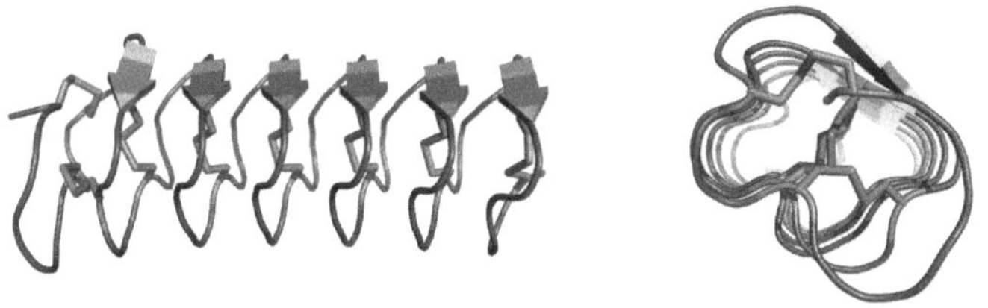

Fig. 7.8 3D presentation of antifreeze protein (1EZG) from two angles. Colors distinguish different fragments of the molecule: solenoid – yellow, disulfides – orange.

Flattened double-walled solenoid

Another solenoid which matches the title of this chapter can be found in a protein listed in PDB under ID 2ZU0: an (E. *coli* sufC-sufD complex involved in the iron-sulfur cluster biosynthesis. The protein is a very large homodimer, with each monomer consisting of two chains. Chain A comprises residues 8–423 (and the same goes for chain B). Chain C has a length of 247 residues and is symmetrical to chain D, although in the latter case only 42 residues are present in PDB file.

Arguably the most interesting component of this protein is the A/B chain dimer, the bulk of which is represented by a large solenoid spanning chain A and continuing in chain B (in a face-on arrangement). The entire complex is characterized by high values of RD and variable correlation coefficients – very low HvT and TvO coupled with very high HvO. This indicates that the listed structure is dominated by intrinsic hydrophobicity (Fig. 7.9).

Overall, the solenoid is structurally similar to amyloids (Fig. 7.10), as evidenced by high values of RD, negative values of both HvT and TvO, and strongly positive values of HvO.

The A/B chain interface is composed of fragments at 355–366 and 340–352 in chain A, along with the corresponding fragments in chain B (antiparallel arrangement). In the context of the solenoid, these fragments are generally accordant with the theoretical distribution, which indicates that hydrophobic interactions play a role in complexation.

Good accordance is also observed for the fragment at 94–105, which functions as a cap. In line with expectations, the outer surface of this fragment faces the environment and mediates entropically advantageous contact with water, preventing unrestricted complexation.

Other fragments exhibit variable deviations from the theoretical distribution, largely resembling an amyloid (the next chapter provides an in-depth

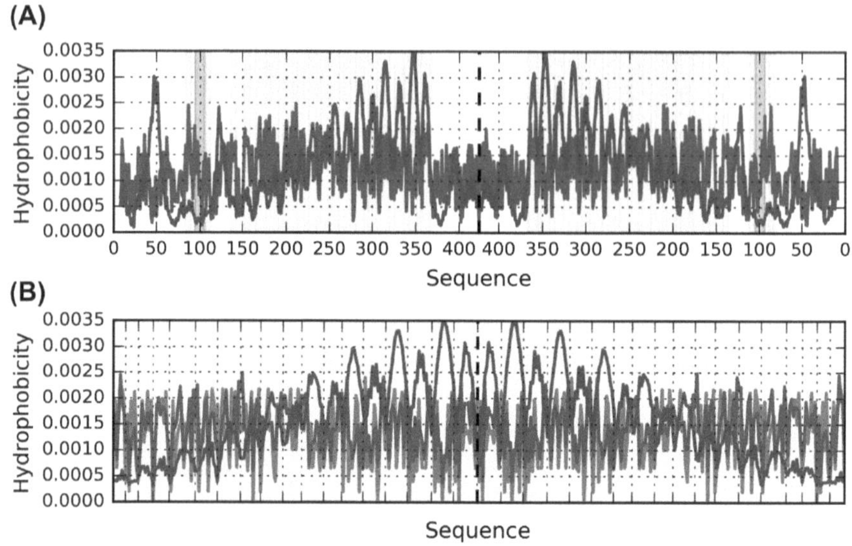

Fig. 7.9 Theoretical (T, blue), observed (O, red) and intrinsic (H, green) hydrophobicity distribution profiles for sufC-sufD complex (2ZU0). (A) Complete molecule, (B) Solenoid fragments (β-strands given in Table 7.4, extended by 1 residue in each direction). Background colors highlight the position of solenoid (yellow) and "stop" fragments (green). Dashed lines separate chain A (left) from chain B (right). To reproduce the symmetry of the complex, results for chain B are present in reverse order. To avoid displaying multiple gaps between the fragments of the solenoid, profiles in B are conflated into contiguous sequence. However ticks on the horizontal axis can be used to determine the location of those fragments withing the complete sequence.

quantitative analysis of amyloid structures in terms of their FOD parameters). In particular, these fragments are dominated by intrinsic hydrophobicity, with high RD coupled with very low values of both TvO and HvT. Such conditions are associated with the propagation of a pattern of alternating bands, as shown in Table 7.4 and Fig. 7.11.

The vast majority of helices observed in chain A (as well as in chain B) remains accordant with the model, suggesting that they play a role in "encapsulating" the solenoid and ensuring that the protein remains soluble. In terms of our model, they act as bridges between the internal (amyloid-like) structure and the external environment, favoring formation of a micelle-like protein, with a predictable hydrophobicity gradient and strongly hydrophilic residues exposed on the surface.

Non-solenoid N-terminal fragment (8—59, 114—134) represents status accordant with idealized distribution while C-terminal fragment

Table 7.4 FOD parameters (RD values and correlation coefficients) for sufC-sufD complex (2ZU0). Results are listed for the complete molecule, for the β-sheets comprising the solenoid, for the solenoid as a whole and for other fragments of the chain, including selected secondary folds. First group of values represents the entire complex as well as each of the four chains analyzed in the context of the complex. "Solenoid (Chain A + Chain B)" corresponds to the status of the solenoid which is formed by chains A and (B).

		RD		Correlation coefficient		
	Fragment	**T-O-R**	**T-O-H**	**HvT**	**TvO**	**HvO**
Complex	A, B, C, D	**0.768**	**0.712**	0.034	0.152	0.754
	Chain A	**0.714**	**0.663**	0.069	0.192	0.762
	Chain B	**0.758**	**0.700**	0.034	0.165	0.754
	Chain C	**0.732**	**0.647**	0.016	0.062	0.793
	Chain D[a]	**0.782**	**0.839**	0.003	-0.126	0.661
Solenoid (Chain A + Chain B)						
β-sheet	**84-89**[b]	**−0.814**	**0.666**	**-0.236**	**-0.112**	**0.830**
β-sheet	**107-111**[b]	**0.831**	**0.781**	**-0.177**	**-0.213**	**0.785**
Complete		**0.804**	**0.703**	**-0.084**	**-0.081**	**0.787**
β-sheet	84–89	0.472	0.735	0.581	0.349	0.818
84-89[b]		0.474	0.772	0.580	0.344	0.841
	152–158	0.686	0.700	-0.438	0.013	0.668
		0.687	0.730	-0.396	-0.019	0.708
	182–191	0.534	0.436	0.154	0.164	0.935
		0.535	0.474	0.156	0.163	0.935
	212–220	0.603	0.242	-0.441	-0.228	0.863
		0.604	0.242	-0.415	-0.207	0.836
	240–247	0.802	0.401	-0.156	-0.401	0.327
		0.791	0.333	-0.181	-0.284	0.121
	267–274	0.507	0.351	0.175	0.250	0.511
		0.528	0.374	0.164	0.172	0.546
	296–305	0.584	0.272	-0.170	-0.284	0.881
		0.587	0.277	-0.175	-0.299	0.888
	326–336	0.615	0.716	-0.088	-0.234	0.894
		0.617	0.697	-0.088	-0.302	0.861
	356–366	0.383	0.215	0.365	0.634	0.929
		0.419	0.247	0.416	0.533	0.972
β-sheet	107–111	0.421	0.305	0.822	0.857	0.878
107-111[b]		0.424	0.253	0.559	0.760	0.785
	137–142	0.585	0.397	0.097	0.345	0.769
		0.650	0.505	0.047	0.102	0.802
	166–177	0.593	0.563	0.158	-0.059	0.686
		0.610	0.565	0.134	-0.114	0.646
	197–207					

(Continued)

Table 7.4 FOD parameters (RD values and correlation coefficients) for sufC-sufD complex (2ZU0). Results are listed for the complete molecule, for the β-sheets comprising the solenoid, for the solenoid as a whole and for other fragments of the chain, including selected secondary folds. First group of values represents the entire complex as well as each of the four chains analyzed in the context of the complex. "Solenoid (Chain A + Chain B)" corresponds to the status of the solenoid which is formed by chains A and (B).—cont'd

		RD		Correlation coefficient		
	Fragment	T-O-R	T-O-H	HvT	TvO	HvO
		0.594	0.539	-0.159	-0.148	0.575
		0.606	0.551	-0.159	-0.172	0.567
	226–235	0.672	0.471	-0.316	-0.382	0.749
		0.686	0.481	-0.327	-0.419	0.756
	252–261	0.635	0.736	-0.631	-0.638	0.839
		0.644	0.734	-0.672	-0.662	0.839
	281–290	0.612	0.526	-0.318	-0.580	0.827
		0.626	0.501	-0.318	-0.625	0.819
	310–319	0.530	0.645	0.216	0.103	0.879
		0.528	0.620	0.215	0.123	0.872
	342–351	0.583	0.333	-0.216	-0.447	0.934
		0.580	0.302	-0.195	-0.368	0.944
Other fragments (CHAIN A)						
Stop – helix + RC	94–105	0.293	0.418	0.427	0.782	0.749
Helices	9–20	0.392	0.291	0.787	0.590	0.565
	26–40	0.419	0.795	0.403	0.528	0.905
	54–59	0.499	0.477	0.632	0.501	0.940
	70–78	0.631	0.577	0.751	0.499	0.774
	95–98	0.334	0.200	0.047	0.711	0.733
	124–134	0.520	0.360	0.040	0.123	0.774
	368–378	0.670	0.665	-0.154	0.113	0.787
	381–399	0.669	0.698	0.317	0.390	0.812
	403–417	0.410	0.227	0.503	0.547	0.938
RC	21–26	0.455	0.527	0.262	0.486	0.917
	39–53	0.424	0.275	0.452	0.466	0.742
	78–90	0.386	0.445	0.538	0.582	0.718
	112–124	0.332	0.255	9,213	0.689	0.739
N-term non-solenoid	8-59 + 114-134	0.394	0.369	0.474	0.769	0.694
C-term non-solenoid	369–423	0.514	0.490	0.300	0.583	0.758

[a]PDB data provides only partial information regarding the structure of chain D.
[b]Numbers correspond to the position of the initial beta fold which belongs to this sheet. Values listed in boldface indicate amyloid-like properties. Underlined values correspond to fragments which "package" the solenoid and are consistent with the fuzzy oil drop model.

3. The solenoid exhibits very strong linear propagation of alternating bands of hydrophobicity, as shown in Fig. 7.11. Negative correlation coefficients indicate that the structure of the presented fragments satisfies amyloid identification criteria. Fig. 7.11 illustrates the peculiar type of discordance exemplified by linear propagation of variably hydrophobic bands in β-sheet starting with strand 107—111 (with short strands omitted). Local minima can be observed in the central part of each strand, where the theoretical model expects hydrophobicity to remain high. Similar distribution is present in the other β-sheet.

(369—423) is discordan (Fig. 7.12) due to a helix (369—377) and a random coil which acts as the P—P interface with chain C and D.

Conclusions and discussion

All of the proteins discussed above exemplify the title of the chapter and each includes fragments characterized by linear propagation of alternating bands of high and low hydrophobicity. While this pattern represents a type of structural ordering, it is unlike the monocentric hydrophobic core expected by the fuzzy oil drop model. When found in biologically active proteins, it calls for a suitable "packaging" which can mediate contact with the aqueous environment (given that the environment promotes a monocentric distribution of hydrophobicity). In principle, linear patterns are capable of unrestricted propagation; however in order to prevent pathologies this capability must be curtailed: thus, solenoid fragments are accompanied by terminal "caps" and — frequently — highly accordant parallel helices, all of which enable entropically advantageous interactions with the solvent (examples may be found by searching PDB for occurrences of "lyase" and "antifreeze" keywords).

In further chapters we will shift our attention to amyloids, which also exhibit band-like linear patterns of hydrophobicity, but lack suitable "packaging" capable of preventing unrestricted growth. These properties will be studied on the example of the beta amyloid (Aβ(1—42) [7,8]) as well as a representative prion amyloid. Similar analysis which focuses on a tau amyloid can be found in Ref. [9].

Clearly, the presence of a solenoid fragment has a profound impact on the structure of the entire protein, and also affects the surrounding aqueous environment. In the case of antifreeze proteins this feedback embodies their biological function, whereas in the presented lyase and the sufC-sufD complex it enables the protein to transmit information to its environment. It is

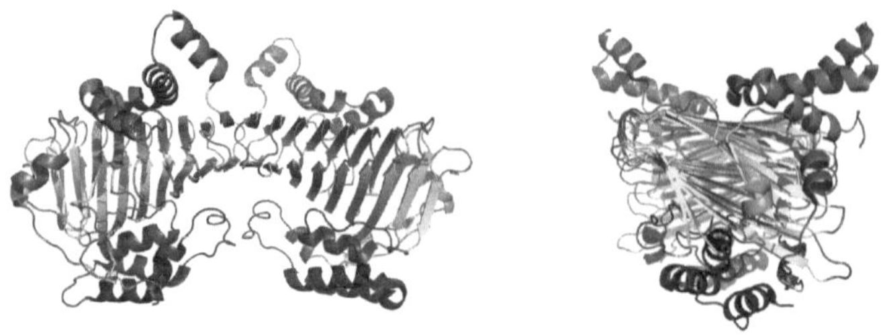

Fig. 7.10 3D presentation of sufC-sufD complex (2ZU0) from two angles. Colors distinguish different fragments of the molecule: solenoid – yellow, "stop" fragments – green, N-terminal non-solenoid structure (8–59, 114–134) – blue, C-terminal non-solenoid structure (369–423) – red.

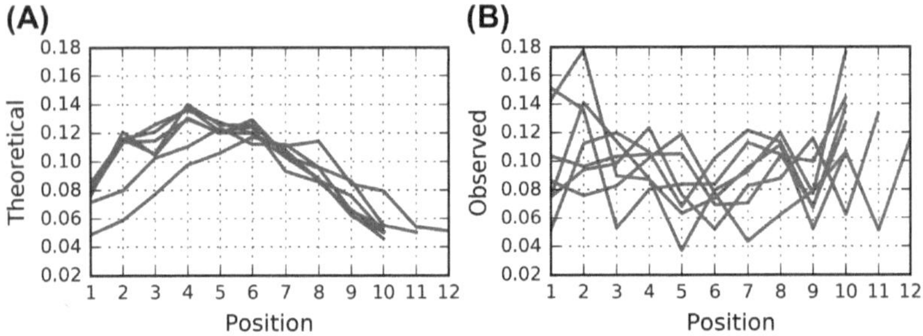

Fig. 7.11 Theoretical (T, blue) and observed (O, red) hydrophobicity distribution profiles for sufC-sufD complex (2ZU0) β-sheet which starts with 107–111 strand. (A) Theoretical hydrophobicity, (B) Observed hydrophobicity.Numbers on horizontal axes denote positions of residues relative to sequence fragments of the β-strands after their alignment with their spatial distribution in the β-sheet.

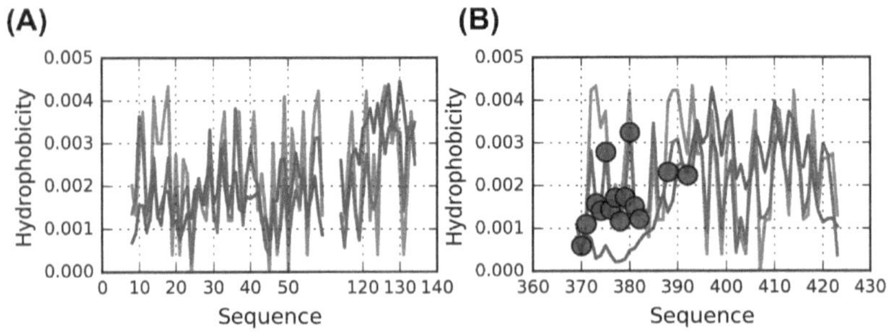

Fig. 7.12 Theoretical (T, blue), observed (O, red) and intrinsic (H, green) hydrophobicity distribution profiles for non-solenoid fragments of chain A of sufC-sufD complex (2ZU0). (A) N-terminal non-solenoid fragment (8–59, 114–134), (B) C-terminal non-solenoid fragment (369–423). Magenta circles mark residues engaged in P–P interaction with chain C.

plausible that the intended ligand is guided to the active site by "sensing" specific local conditions, which are generated by the protein. Such signaling may be regarded as equally important as the influence of the solvent upon the folding process (promoting formation of spherical micelles), while the internal force field — primarily determined by the presence of the solenoid fragment — facilitates catalysis.

The solenoids presented in this chapter may be thought of as ribbonlike micelles. In the Chapter 3. devoted to information theory we suggested that both spherical and ribbonlike micelles carry very little information (which is attributable to their predictable symmetry). Nevertheless, the abovementioned proteins may be regarded as information-rich due to the presence of packaging which encapsulates the solenoid. In contrast, amyloids — devoid of any such packaging — are essentially bare micelles and therefore contain no information. This issue will be further discussed in subsequent chapters.

References

[1] Do H, Kim S-J, Kim HJ, Lee JH. Structure-based characterization and antifreeze properties of a hyperactive ice-binding protein from the Antarctic bacterium Flavobacterium frigorisPS1. Acta Crystallographica Section D Biological Crystallography 2014;70(4): 1061–73. https://doi.org/10.1107/s1399004714000996.

[2] Schutzius TM, Jung S, Maitra T, Graeber G, Köhme M, Poulikakos D. Spontaneous droplet trampolining on rigid superhydrophobic surfaces. Nature 2015;527(7576): 82–5. https://doi.org/10.1038/nature15738.

[3] Macias-Romero C, Nahalka I, Okur HI, Roke S. Optical imaging of surface chemistry and dynamics in confinement. Science 2017;357(6353):784–8. https://doi.org/10.1126/science.aal4346.

[4] Banach M, Konieczny L, Roterman I. Why do antifreeze proteins require a solenoid? Biochimie 2018;144:74–84. https://doi.org/10.1016/j.biochi.2017.10.011.

[5] Pickersgill R, Jenkins J, Harris G, Nasser W, Robert-Baudouy J. The structure of Bacillus subtilis pectate lyase in complex with calcium. Nature Structural & Molecular Biology 1994;1(10):717–23. https://doi.org/10.1038/nsb1094-717.

[6] Liou Y-C, Tocilj A, Davies PL, Jia Z. Mimicry of ice structure by surface hydroxyls and water of a β-helix antifreeze protein. Nature 2000;406(6793):322–4. https://doi.org/10.1038/35018604.

[7] Dułak D, Banach M, Gadzała M, Konieczny L, Roterman I. Structural analysis of the Aβ(15-40) amyloid fibril based on hydrophobicity distribution. Acta Biochimica Polonica 2018. https://doi.org/10.18388/abp.2018_2647.

[8] Banach M, Konieczny L, Roterman I. Fuzzy oil drop model application—from globular proteins to amyloids. Computational Methods to Study the Structure and Dynamics of Biomolecules and Biomolecular Processes 2018:639–58. https://doi.org/10.1007/978-3-319-95843-9_1.

[9] Dułak D, Gadzała M, Banach M, Ptak M, Wiśniowski Z, Konieczny L, Roterman I. Filamentous aggregates of tau proteins fulfil standard amyloid criteria provided by the fuzzy oil drop (FOD) model. International Journal of Molecular Sciences 2018; 19(10):2910. https://doi.org/10.3390/ijms19102910.

CHAPTER 8

Composite structures

Mateusz Banach[1], Leszek Konieczny[2], Irena Roterman[1]
[1]Department of Bioinformatics and Telemedicine, Jagiellonian University — Medical College, Krakow, Poland
[2]Chair of Medical Biochemistry, Jagiellonian University — Medical College, Krakow, Poland

Contents

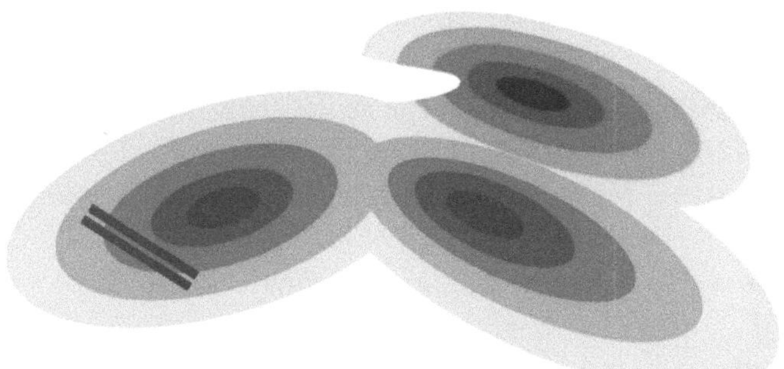

Conceptual diagram illustrating a large, complex protein with several roles (p-p interaction, ligand binding cavity), along with fragments characterized by linear distribution of hydrophobicity. In such a protein the active site may be constructed from multiple structural units, each of which — when considered on its own — follows the micellar pattern.

Analysis of a nonredundant set of PDB proteins reveals that a vast majority of individual domains exhibit strong micelle-like characteristics [1]. A

From Globular Proteins to Amyloids
ISBN: 978-0-08-102981-7
https://doi.org/10.1016/B978-0-08-102981-7.00011-7

large protein comprising several chains which are further subdivided into domains, may be regarded as a composite structure, where each unit contributes some of the required information. Multiple units may cooperate to form an active site characterized by specific deviations from the monocentric distribution of hydrophobicity. In certain cases, such active sites are not only discordant but in active opposition to the monocentric distribution – for example, they may exhibit a linear pattern of alternating bands of high and low hydrophobicity.

Single-chian enzyme consisting of three domains

The first example of a composite structure is provided by a single-chain enzyme: hydrolase (E.C. 3.1.5.1) elongation factor Ef-tu complexed with a GTP analog in the antibiotic pulvomycin [2] (structure available as 2C78 in PDB).

The protein catalyzes the following reaction: dGTP + H_2O => deoxyguanosine + triphosphate.It consists of three domains, with catalytic residues located in domain 1.

Analysis of the molecule as a whole

The structure as a whole is not globular and does not contain a prominent hydrophobic core (Table 8.1). This particular protein interacts with two ligands:

GNP – Phosphoaminophosphonic acid-Guanylate ester.

PUL – *(1s,2s,3e,5e,7e,10s,11s,12s)-12-[(2r,4e,6e,8z,10r,12e, 14e,16z, 18s,19z)-10,18-Dihydroxy-12,16,19-Trimethyl- 11,22-Dioxooxacyclodocosa-4, 6,8,12,14,16,19-Heptaen-2- Yl]-2,11-Dihydroxy-1,10-Dimethyl-9-Oxotrideca-3,5,7- Trien-1-Yl 6-Deoxy-2,4-Di-O-Methyl-Beta-L- Galactopyranoside [Pulvomycin]*

Elimination of ligand binding residues lowers the value of RD, which indicates that the binding pocket contains information (although note that even under these conditions RD remains greater than 0.5, suggesting that other factors are at play in terms of producing deviations between O and T).

Elimination of catalytic residues produces a further decrease in RD, which, again, shows that these residues contain information required in the process of catalysis. When considering these residues we also take into account their immediate neighborhood (5 aa in each direction). Notably, 85H is the only catalytic residue whose neighborhood fragment significantly diverges from the theoretical distribution.

Table 8.1 Values of fuzzy oil drop parameters calculated for the structure of elongation factor Ef-tu (2C78) and its three domains. Values listed as boldfaces correspond to fragments of status discordant versus the idealized one; values underlined – fragments representing highly ordered status in respect to idealized distribution.

		RD		Correlation coefficient		
Hydrolase (2C78)	**Fragment**	**T-O-R**	**T-O-H**	**HvT**	**TvO**	**HvO**
	9–405	**0.627**	**0.593**	0.211	0.385	0.773
Lig		**0.746**	**0.502**	**−0.665**	**−0.005**	**0.647**
No lig		**0.591**	**0.584**	0.260	0.463	0.788
No cat res	21D, 85H	**0.618**	**0.584**	**0.210**	**0.385**	**0.772**
Cat 21D	16–26	0.464	0.464	0.201	0.606	0.786
Cat 85H	79–90	0.465	0.465	0.234	0.490	0.670
Domain 1	9–213	**0.594**	**0.571**	**0.294**	**0.497**	**0.766**
Domain 2	216–308	**0.612**	**0.587**	**0.092**	**0.254**	**0.800**
Domain 3	311–404	**0.637**	**0.569**	**0.120**	**0.311**	**0.775**
Secondary structure						
β-strands	10–17	0.198	0.279	0.336	0.849	0.647
	66–72	0.496	0.213	0.383	0.845	0.722
	75–82	0.228	0.358	0.520	0.586	0.844
	102–108	**0.979**	**0.753**	0.507	0.758	0.533
	130–136	0.626	0.534	0.522	0.309	0.652
	169–173	0.361	0.433	0.446	0.648	0.842
β-sheet		**0.505**	**0.554**	0.342	0.559	0.691
Helices	23–38	**0.698**	0.427	**−0.142**	**0.042**	**0.777**
	46–51	0.395	0.295	0.500	0.648	0.826
	53–60	0.369	0.129	−0.200	0.773	0.371
	85–88	0.388	0.225	0.769	0.597	0.965
	89–97	0.362	0.292	0.402	0.913	0.586
	113–126	**0.556**	**0.516**	**0.303**	**0.433**	**0.881**
	137–141	**0.687**	0.214	**−0.587**	**−0.269**	**0.789**
	143–161	0.643	0.660	0.308	0.408	0.887
	174–185	0.234	0.299	0.453	0.817	0.722
	193–210	0.463	0.397	0.535	0.592	0.924
β-strand	**223–225**	**0.679**	**0.662**	**−0.992**	**−0.974**	**0.940**
	227–231	**0.832**	**0.764**	**−0.489**	**−0.335**	**0.898**
	235–241	**0.886**	**0.984**	**−0.545**	**−0.584**	**0.938**
	246–248	0.175	0.008	0.801	0.900	0.982
	252–256	0.144	0.154	0.743	0.916	0.884
	263–272	**0.603**	**0.716**	**0.257**	**0.013**	**0.822**
	275–277	**0.574**	0.140	**−0.668**	**−0.224**	**0.876**
	279–281	0.169	0.348	0.984	0.997	0.994
	285–290	**0.837**	**0.952**	**−0.749**	**−0.789**	**0.851**

(Continued)

Table 8.1 Values of fuzzy oil drop parameters calculated for the structure of elongation factor Ef-tu (2C78) and its three domains. Values listed as boldfaces correspond to fragments of status discordant versus the idealized one; values underlined – fragments representing highly ordered status in respect to idealized distribution.—cont'd

		RD		Correlation coefficient		
Hydrolase (2C78)	**Fragment**	**T-O-R**	**T-O-H**	**HvT**	**TvO**	**HvO**
β sheet		0.664	0.560	0.660	0.143	0.800
β-strands	311–322	0.238	0.088	0.672	0.812	0.832
	341–343	0.588	0.627	0.528	−0.001	0.848
	347–354	**0.787**	**0.905**	**−0.222**	**−0.338**	**0.871**
	367–380	0.393	0.206	0.387	0.611	0.847
	385–389	**0.529**	**0.945**	**−0.101**	**−0.058**	**0.965**
	393–403	0.248	0.132	0.235	0.807	0.604
β-sheet		0.573	0.422	0.218	0.394	0.777
Helix	324–327	0.448	0.034	0.712	0.587	0.018
Domain A1 – individual						
	9–213	0.394	0.378	0.395	0.695	0.767
Lig. binding		**0.652**	0.303	**−0.228**	**0.221**	**0.648**
No lig. bind.		0.352	0.399	0.487	0.750	0.784
No E		0.390	0.372	0.400	0.700	0.765
Cat E21	61–26	0.553	0.222	−0.187	0.176	0.766
Cat 85H	79–90	0.350	0.218	0.224	0.785	0.669
Secondary structure						
β-strand	10–17	0.255	0.350	0.328	0.853	0.647
	66–72	0.263	0.201	0.187	0.772	0.722
	75–82	0.307	0.247	0.776	0.747	0.843
	102–108	0.946	0.536	0.759	0.795	0.528
	130–136	0.424	0.335	0.132	0.574	0.652
	169–173	0.253	0.317	0.739	0.937	0.842
β-sheet		0.284	0.326	0.463	0.791	0.691
Helices	**23–38**	**0.672**	**0.397**	**0.052**	**0.191**	**0.777**
	46–51	0.426	0.322	0.687	0.558	0.825
	53–60	0.445	0.170	0.070	0.668	0.368
	85–88	0.034	0.016	0.895	0.979	0.966
	89–97	0.224	0.173	0.435	0.894	0.585
	113–126	0.517	0.642	0.262	0.420	0.886
	137–141	**0.707**	0.231	**−0.660**	**−0.347**	**0.789**
	143–161	0.412	0.429	0.411	0.626	0.886
	174–185	0.263	0.330	0.428	0.814	0.720
	193–210	0.391	0.328	0.495	0.623	0.924

Table 8.1 Values of fuzzy oil drop parameters calculated for the structure of elongation factor Ef-tu (2C78) and its three domains. Values listed as boldfaces correspond to fragments of status discordant versus the idealized one; values underlined – fragments representing highly ordered status in respect to idealized distribution.—cont'd

		RD		Correlation coefficient		
Hydrolase (2C78)	**Fragment**	**T-O-R**	**T-O-H**	**HvT**	**TvO**	**HvO**
Domain A2 – individual						
	216–307	0.310	0.296	0.515	0.770	0.801
Lig. binding		0.309	0.307	0.158	0.753	0.710
No lig. bind		0.310	0.296	0.562	0.772	0.820
Secondary structure						
β-strand	223–225	0.178	0.258	0.684	0.928	0.906
	227–231	0.434	0.333	0.377	0.527	0.897
	235–241	0.234	0.734	0.633	0.834	0.937
	246–248	0.069	0.003	0.990	0.998	0.981
	252–256	0.320	0.306	0.723	0.722	0.957
	263–272	0.363	0.484	0.398	0.743	0.822
	275–277	0.195	0.028	0.925	0.992	0.875
	279–281	0.453	0.678	0.300	0.407	0.993
	285–290	0.170	0.446	0.781	0.880	0.851
β-sheet		0.302	0.217	0.502	0.770	0.799
Domain A3 – individual						
	311–404	0.340	0.298	0.504	0.748	0.767
Lig. binding		0.071	0.117	0.605	0.976	0.635
No ligand		0.342	0.288	0.499	0.751	0.783
Secondary structure						
β-strands	311–322	0.434	0.236	0.447	0.719	0.784
	341–343	0.052	0.061	0.854	0.999	0.849
	347–354	0.314	0.529	0.613	0.756	0.841
	367–380	0.437	0.311	0.473	0.698	0.893
	385–389	0.083	0.489	0.880	0.973	0.954
	393–403	0.286	0.155	0.307	0.793	0.609
β-sheet		0.367	0.277	0.499	0.736	0.771
Helix	324–327	0.272	0.022	−0.321	0.818	−0.348

Going back to the molecule as a whole, we may note that none of its constituent domains are consistent with the theoretical distribution (when analyzed as components of the protein). The status of secondary and super-secondary folds varies; however, all β-sheets are strongly discordant.

Domains treated as independent structural units

In contrast to the above, when considered as standalone units, all three domains exhibit strong accordance with the monocentric core pattern. We will attempt to provide an explanation for this phenomenon.

The final structure, which — for environmental reasons (among others) — must exhibit a nonstandard distribution of hydrophobicity, is formed by joining together three individual components, each of which has been shaped by hydrophobic interactions and resembles a micelle. Even domain 1, which contains the protein's catalytic residues, is regarded as accordant (note that catalytic residues are expected to carry information and therefore exhibit deviations from the Gaussian distribution of hydrophobicity).

The structure also includes ligands, one of which acts as an inhibitor. From the point of view of domains 2 and 3, the status of ligand-binding residues remains highly accordant with the theoretical distribution. This suggests that a structure capable of recognizing and binding the correct ligand emerges as a result of interactions between several independent domains.

The status of catalytic residues (found only in domain 1) varies depending on whether we consider the domain by itself or the molecule as a whole. While a specific deviation has been produced inside the domain for residue H85 ($\pm$5 aa), residue E21 does not disrupt its overall micellar pattern. Its status changes (to discordant) only after the respective domains assemble into a composite protein.

The status of catalytic residues can be determined by calculating FOD parameters for the input chain following elimination of such residues. The value of RD decreases in the process, however, it remains greater than 0.5.

Plotting hydrophobicity distribution charts for the entire chain (Fig. 8.1) reveals multiple deviations, particularly in the central area where hydrophobicity is much lower than expected. The presented chart also shows interdomain boundaries. This situation changes when domains are analyzed as standalone units. In particular, Fig. 8.2 reveals strong accordance between T and O for each domain.

The presented analysis suggests that catalytically active structures may be produced by assembling simple micellar components whose O is aligned

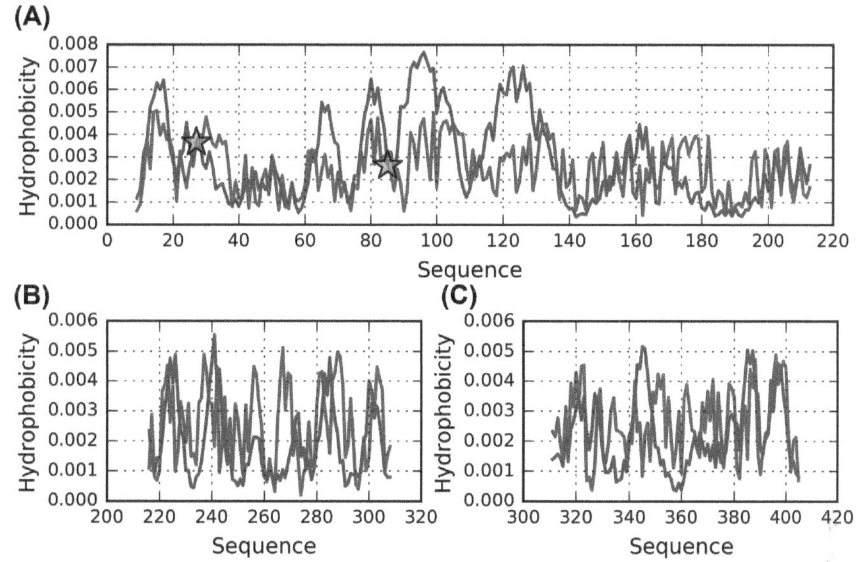

Fig. 8.1 Theoretical (T, blue) and observed (O, red) hydrophobicity distribution profiles for three domains of 2C78 treated as whole protein. (A) domain 1 (9–213), (B) domain 2 (216–307), (C) domain 3 (311–404) Orange stars mark positions of catalytic residues.

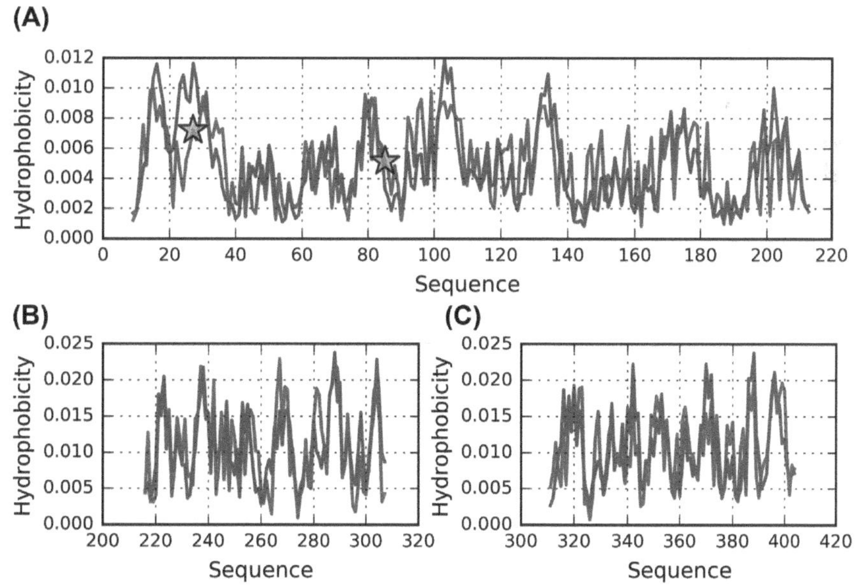

Fig. 8.2 Theoretical (T, blue) and observed (O, red) hydrophobicity distribution profiles for three domains of 2C78 treated as individual units. (A) domain 1 (9–213), (B) domain 2 (216–307), (C) domain 3 (311–404) Orange stars mark positions of catalytic residues.

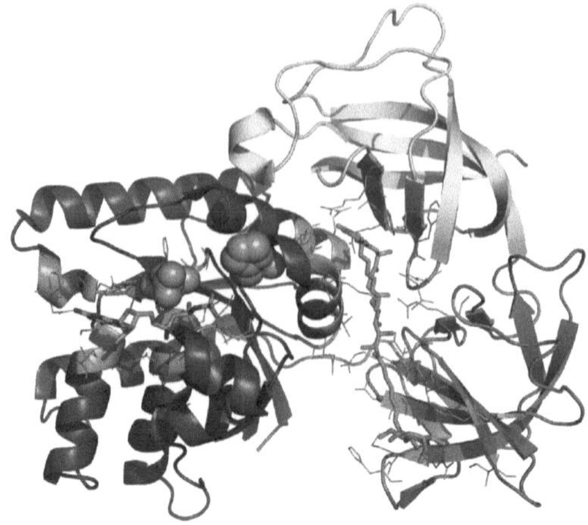

Fig. 8.3 3D presentation of 2C78, with colors distinguishing its three domains: (A) 1 – blue, (B) 2 – red, (C) 3 – yellow. Residues engaged in protein-ligand interactions are green-colored and have side chains displayed. Orange spheres mark catalytic residues (21D, 85H).

with T. The assembly process by itself introduces specific deviations from the monocentric distribution of hydrophobicity, facilitating biological function (Fig. 8.3).

Homodimer enzyme

The second sample protein analyzed in this chapter is the isozyme of citrate synthase from *Sulfolbus tokodaii* strain 7 (PDB ID: 1VGM) [3]. This enzyme is classified as E.C. 2.3.3.1 and participates in the following reaction:

citrate (Si)-synthase – Acetyl-CoA + H_2O + oxaloacetate ⇒ citrate + CoA

The protein is a homodimer (CATH classification: 5.1.1.1), with complex secondary and supersecondary characteristics (Table 8.2). Each monomeric unit is composed of two domains. Structural assessment indicates the lack of a monocentric hydrophobic core (Table 8.2, Figs. 8.4 and 8.5). Elimination of interface residues results in only a slight decrease in RD. Thus, the interface is not thought to significantly disrupt the distribution of hydrophobicity in the dimer. Similarly, elimination of catalytic residues (as well as their immediate neighborhood – 5 aa in each direction) does not produce a marked change in the status of the molecule. Regarding the catalytic residues themselves, their status is highly variable. Residue 189S (together with its neighborhood) conforms to the theoretical distribution,

Table 8.2 Values of fuzzy oil drop parameters calculated for the structure of 1VGM and its two domans. Values listed in boldface indicate conditions which contrast with a micelle-like conformation. Double entries in the top section of the table correspond to chains A and B respectively. "No arms" indicates chains A and B following elimination of "arms" which protrude from their globular structures (fragments 3–13 and 356–378 respectively).

		RD		Correlation coefficient		
Isozyme of citrate synthase (1VGM)	**Fragment**	**T-O-R**	**T-O-H**	**HvT**	**TvO**	**HvO**
Dimer						
	2 × (3–378)	**0.622**	**0.553**	0.236	0.556	0.736
P-P		**0.604**	**0.531**	0.093	0.600	0.640
No P-P		**0.615**	**0.550**	0.291	0.551	0.760
No E	Eliminated: 189S, 219H 259H, 314D	**0.619**	**0.553**	0.240	0.556	0.739
No E	Enzymatic residues ± 5aa	**0.611**	**0.553**	0.277	0.564	0.758
189S	184–194	0.210	0.214	0.190	0.955	0.405
		0.200	0.195	0.177	0.944	0.409
219H	214–224	**0.579**	0.176	0.095	0.099	0.733
		0.580	0.185	0.101	0.086	0.727
259H	254–264	**0.528**	0.361	0.392	0.243	0.768
		0.353	0.277	0.532	0.630	0.681
314D	309–319	**0.823**	**0.700**	**−0.617**	**−0.343**	**0.642**
		0.805	**0.700**	**−0.672**	**−0.484**	**0.697**
B-sheet	19-23, 28–32, 34-35	**0.761**	0.488	0.283	0.196	0.613
		0.768	0.479	0.328	0.289	0.607
β-sheet	A 12–22 B 359–369	**0.564**	**0.804**	**0.046**	**0.179**	**0.924**
C-term	368–378	0.494	0.296	0.195	0.290	0.771
		0.436	0.225	0.311	0.547	0.794

(Continued)

Table 8.2 Values of fuzzy oil drop parameters calculated for the structure of 1VGM and its two domans. Values listed in boldface indicate conditions which contrast with a micelle-like conformation. Double entries in the top section of the table correspond to chains A and B respectively. "No arms" indicates chains A and B following elimination of "arms" which protrude from their globular structures (fragments 3–13 and 356–378 respectively).—cont'd

		RD		Correlation coefficient		
Isozyme of citrate synthase (1VGM)	**Fragment**	**T-O-R**	**T-O-H**	**HvT**	**TvO**	**HvO**
Chain A						
	3–378	**0.550**	0.480	0.253	0.518	0.720
P-P		**0.571**	0.488	0.030	0.454	0.527
No P-P		**0.553**	0.484	0.305	0.519	0.772
189S	184–194	**0.516**	0.494	−0.323	0.008	0.319
219H	214–224	0.438	0.150	0.447	0.610	0.538
259H	254–264	0.447	0.295	0.483	0.453	0.768
314D	309–319	**0.813**	**0.685**	−0.550	−0.271	0.623
β-sheet	19-23, 28–32, 34–35	**0.684**	0.436	0.238	−0.105	0.500
Chain B						
	3–378	**0.534**	0.465	0.260	0.528	0.715
P-P		**0.558**	0.472	0.005	0.454	0.531
No P-P		**0.524**	0.468	0.320	0.532	0.764
189S	184–194	**0.513**	0.493	−0.316	0.077	0.308
219H	214–224	0.444	0.159	0.506	0.572	0.544
259H	254–264	0.300	0.232	0.564	0.740	0.681
314D	309–319	**0.785**	**0.672**	**−0.607**	**−0.446**	**0.696**
β-sheet	19-23, 28–32, 34-35	**0.628**	0.378	0.277	0.103	0.492
Chain A – no arms						
	14–355	**0.584**	**0.502**	0.284	0.453	0.746
P-P		**0.665**	0.487	0.001	0.184	0.498

No P-P		**0.566**	0.499	0.315	0.471	0.780
189S	184–194	**0.575**	0.444	0.294	0.638	0.286
219H	214–224	0.295	0.087	0.200	0.785	0.536
259H	254–264	0.401	0.258	0.530	0.583	0.768
314D	309–319	**0.822**	**0.700**	**-0.524**	**-0.197**	**0.623**
β-sheet	19-23, 28–32, 34-35	**0.743**	**0.560**	0.204	-0.055	0.414
Chain (B) no arms						
	14–355	**0.582**	**0.501**	0.286	0.460	0.738
P-P		**0.678**	**0.504**	-0.066	0.160	0.510
No P-P		**0.562**	0.495	0.324	0.481	0.770
189S	184–194	**0.654**	**0.635**	**-0.402**	**-0.544**	**0.307**
219H	214–224	0.303	0.093	0.219	0.784	0.546
259H	254–264	0.237	0.180	0.578	0.813	0.682
314D	309–319	**0.805**	**0.700**	**-0.574**	**-0.367**	**0.697**
β-sheet	19-23, 28–32, 34-35	**0.734**	**0.576**	0.305	0.166	0.493
Domain A1						
	15-220, 325-358	0.490	0.433	0.329	0.569	0.689
189S	184–194	0.483	0.459	-0.104	0.292	0.304
219H	214–224	0.333	0.093	0.268	0.786	0.577
β-sheet	19-23, 28–32, 34-35	**0.730**	0.463	0.249	0.106	0.452
Domain A2						
	221–324	0.470	0.339	0.335	0.527	0.820
259H	254–264	0.359	0.324	0.536	0.891	0.733
314D	309–319	**0.758**	**0.609**	0.122	-0.192	0.697

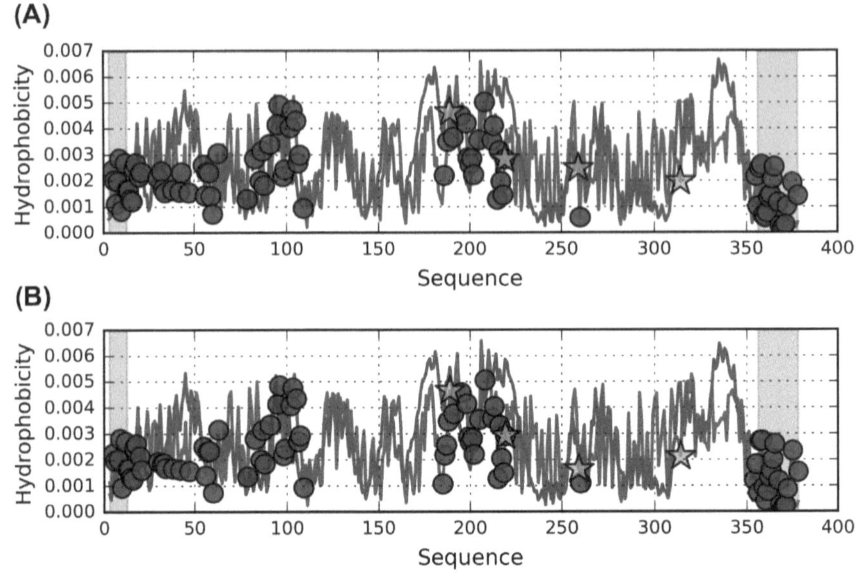

Fig. 8.4 Theoretical (T, blue) and observed (O, red) hydrophobicity distribution profiles for 1VGM. (A) chain A, (B) chain B Magenta circles mark residues engaged in P-P interaction, while stars correspond to catalytic residues. Red background distinguish the location of "arms" protruding from the molecule (fragments 3–13 and 356–378).

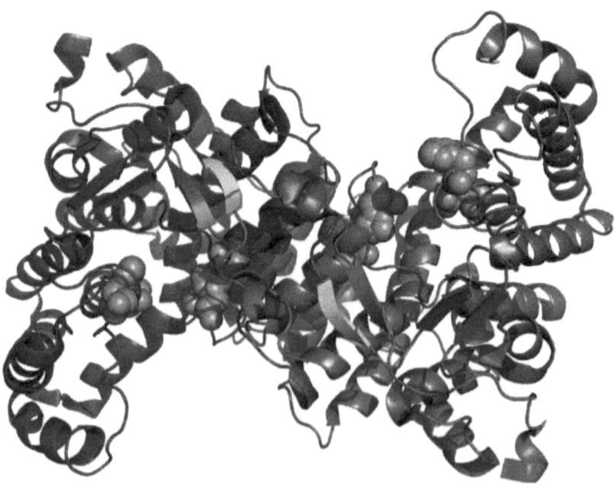

Fig. 8.5 3D presentation of isozyme of citrate synthase (1VGM) homodimer (chain A – blue, chain B – gray). Red fragments in chain A and brown in chain B are "arms" with which the subunits embrace. Yellow fragments (which do not belong to the "arms" themselves) form β-sheets with strands from the "arms". Orange spheres – catalytic residues (189S, 219H, 259H, 314D).

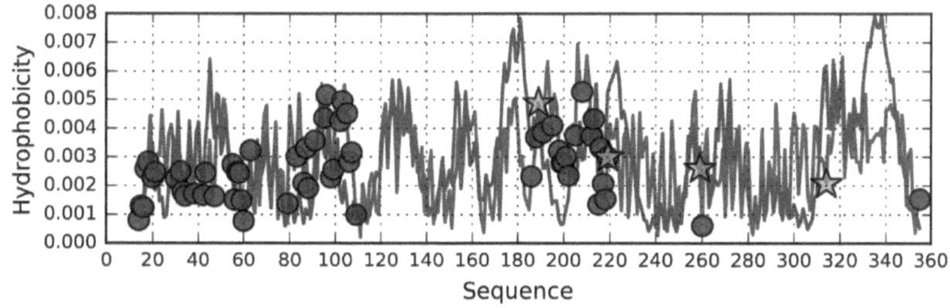

Fig. 8.6 Theoretical (T, blue) and observed (O, red) hydrophobicity distribution profiles for 1VGM chain A without its "arms" – fragment 314–355 (following elimination of fragments 3–13 and 356–378). Magenta circles mark residues engaged in P-P interaction, while stars correspond to catalytic residues.

while residues 219H and 259H diverge from it. 314D, a catalytic residue, actively opposes the theoretical distribution: the conformation of its neighborhood is determined by intrinsic hydrophobicity. Such conditions may have a trickle-down effect on the properties of the protein's aqueous environment.

Strong discordance is also observed for the protein's β-sheets – again, their purpose may be to cause specific structural changes in the solvent.

An interesting structural motif is present in the dimer: two short N-terminal (3–13) and two long C-terminal (356–378) sections ("arms") (Fig. 8.6) protrude from each chian and appear to embrace the other chain. This two-chain structure also includes two antiparallel discordant β-sheets which consist of fragments contributed by both chains: residues 12–22 (chain A) linked with residues 359–369 (chain B) – and *vice versa.*

It should be noted that these fragments – along with the previously mentioned β-sheet – are all found at the "entrance" to the active site (catalytic residues S189, H219, H259, D314). Consequently, they may be suspected of altering the structure of the aqueous environment in order to facilitate the protein's biological function.

Analysis of chains and domains regarded as standalone structures

As already mentioned, the structure of each individual chain consists of a large globular portion and an elongated arm which forms a peculiar "interface". Given that the protruding fragment may significantly hamper FOD analysis of the molecule, it has been excluded from calculations.

Results shown in Table 8.2 reveal an increase in discordance following excision of the protruding "arms". This shows that information is encoded

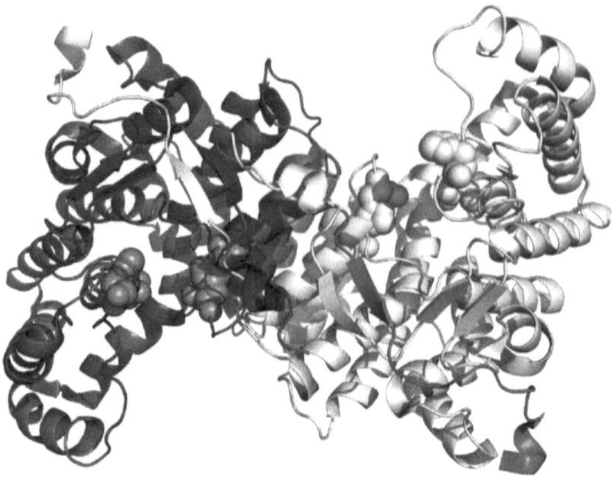

Fig. 8.7 3D presentation of 1VGM homodimer highlighting fragments which conform to a micellar distribution (14–79, 89–100, 109–174, 188–190, 203–205, 211–224, 229–235 and 250–260) – blue. They are shown only in chain A for readability (whole chain B is shown in white). Like on Fig. 8.5, "arm" of chain A is colored red and its four catalytic residues are marked by the orange spheres.

in the globular portion itself. However 'arms" by themselves represent status accordant with expected hydrophobicity distribution. The effect is particularly pronounced for the β-sheet and the catalytic residue at 189S, along with its immediate neighborhood.

Given the complex structure of the enzyme, it is interesting to seek fragments responsible for maintaining its micelle-like form. Strongly micellar fragments include those at 14–79, 89–100, 109–174, 188–190, 203–205, 211–224, 229–235 and 250–260. Taken together, their RD(T-O-R) and (T-O-H) values are 0.454 and 0.348 respectively, with correlation coefficients calculated as 0.352, 0.589 and 0.757 (HvT, TvO and HvO respectively – see Fig. 8.7). They provide the structural backbone which enables the protein to remain stable. Accordance between T and O is evident in Fig. 8.4, particularly for the 109–174 fragment. Note that in order to reduce RD we have eliminated those residues which exhibits major differences between T and O.

Domains

As already remarked, each domain considered on its own (i.e. constructing a 3D Gaussian capsule specifically for that domain) is seen as consistent with the theoretical distribution of hydrophobicity (Table 8.2).

Doman 1, when stripped of ligand-binding residues, reveals a decrease in RD, which indicates that these residues carry information. In some cases,

HvT correlation coefficients calculated for ligand-binding residues adopt negative values, suggesting very strong discordance.

Domain 2 shows excellent accordance with the monocentric distribution and is only slightly disrupted by the presence of the ligand. This accordance is observed both in the neighborhood of ligand-binding residues and in other fragments comprising the domain.

The status of each catalytic residue (along with its 5 aa neighborhood) is as follows:

189S: accordant in the dimer but discordant in the monomer and in the "no arms" structure

219H: discordant in the dimer but accordant in the monomer (including in the "no arms" structure) and in its individual domain

259H: depends on the chain (differing status in chains A and B forming the dimer) but accordant in each individual chain and domain

314D: discordant in all structures subjected to analysis

The above results show that domains, folding individually, shape the neighborhoods of residues 189S, 219H and 259H in accordance with a micelle-like pattern of hydrophobicity. On the other hand, residue 314D

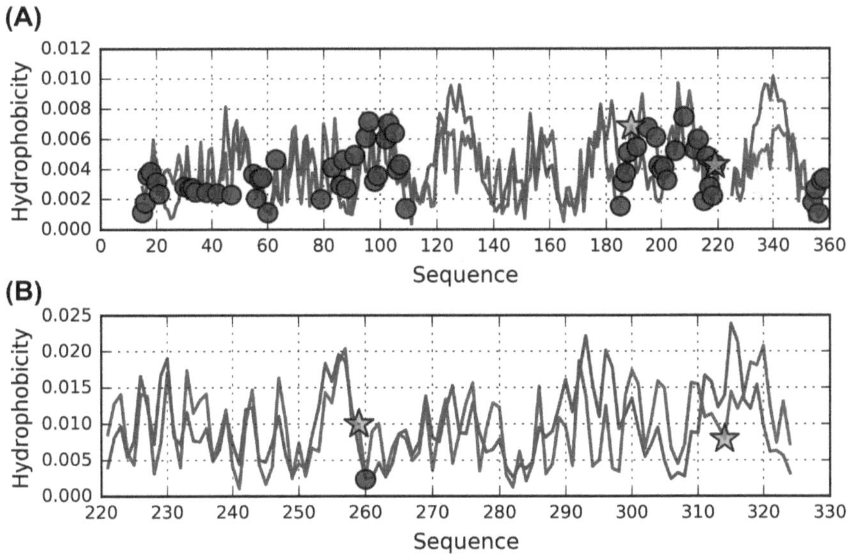

Fig. 8.8 Theoretical (T, blue) and observed (O, red) hydrophobicity distribution profiles for two domains of 1VGM treated as individual units. (A) domain 1 (15−220 + 325−358), (B) domain 2 (221−324) Magenta circles mark residues engaged in P-P interaction, while stars correspond to catalytic residues.

consistently opposes this pattern regardless of the scope of our analysis. This residue therefore carries information by itself, whereas the status of other catalytic residues is determined by interactions between chains forming the dimer (Fig. 8.8).

Conclusions and discussion

In summarizing this part of our analysis, we may conclude that the molecule as a whole differs – somewhat unexpectedly – from a regular micelle, even though environmental factors promote the formation of micellar structures. The protein gains its structural properties in the process of assembly of individual components, each of which folds separately and retains accordance with the theoretical distribution of hydrophobicity (note that the definition of a protein domain includes the requirement for standalone folding). Thus, the information which the protein requires for its biological activity is encoded in the steric arrangement of its constituent domains. The assembly process causes the catalytic active site to diverge from the theoretical distribution.

This conclusion is based on our initial assumption that each protein must fulfill a specific biological role. In the case of enzymes, specific environmental conditions must be produced in order to enable catalysis.

The presented protein is also an example of information encoding through composition. The desired result is achieved by arranging individual components, each of which conforms to the model. When analyzing individual domains on their own, we find that β-structural fragments (β-sheets) conform to the theoretical distribution of hydrophobicity which asserts the presence of a centralized core, and that helical fragments are also consistent with this pattern. If, however, the same secondary folds are analyzed in the context of the protein complex, many of them are found to deviate from the theoretical distribution – consequently, the protein as a whole lacks a clearly defined hydrophobic core.

When attempting to interpret the status of this composite enzyme in light of our introductory hypothesis, we may conclude that – in this case – information is encoded as a specific arrangement of distinct domains, none of which carries much information by itself. The assembly process introduces local deviations with respect to the molecule-wide Gaussian, effectively "imbuing" the protein with information. We may even point to specific places where such information is encoded.

Of course, there many ways in which individual structural units (chains, domains) can assemble into larger structures. The analysis presented in this chapter aims to shed light on the assembly process, and on the structural properties of the resulting composite proteins.

An example of a protein whose complexity is mainly due to secondary structural variations is the lyase discussed in chapter 7. In addition to classic secondary folds, this protein also contains areas characterized by alternating bands of high and low hydrophobicity.

Another highly complex proteins subjected to FOD analysis is the GroEl chaperonin, which accompanies other proteins at the folding stage. This complex may be regarded as a true "molecular machine" given the complexity of its mechanism of action [4]. Its structure, consisting of 21 chains in three layers with a heptagonal symmetry, undergoes such strong activity-related deformations [5] that its symmetry is effectively erased. The base symmetry is likely required for the complex to revert to its original form once its activity cycle has been completed.

F-actin provides another example of a molecular machine due to its structural and functional complexity. This protein is subject to further analysis in Ref. [6], focusing on local structural patterns and the presence of alternating bands of variable hydrophobicity.

References

[1] Sałapa K, Kalinowska B, Jadczyk T, Roterman I. Measurement of hydrophobicity distribution in proteins — non-redundant protein data bank. Bio-Algorithms and Med-Systems 2012;8. https://doi.org/10.2478/bams-2012-0023.

[2] Parmeggiani A, Krab IM, Okamura S, Nielsen RC, Nyborg J, Nissen P. Structural basis of the action of pulvomycin and GE2270 a on elongation factor tu. Biochemistry 2006; 45(22):6846—57. https://doi.org/10.1021/bi0525122.

[3] Murakami M, Kouyama T. Crystal structures of two isozymes of citrate synthase from sulfolobus tokodaii strain 7. Biochemistry Research International 2016:1—11. https://doi.org/10.1155/2016/7560919.

[4] Banach M, Stąpor K, Roterman I. Chaperonin structure - the large multi-subunit protein complex. International Journal of Molecular Sciences 2009;10(3):844—61. https://doi.org/10.3390/ijms10030844.

[5] Xu Z, Horwich AL, Sigler PB. The crystal structure of the asymmetric GroEL—GroES—(ADP)7 chaperonin complex. Nature 1997;388(6644):741—50. https://doi.org/10.1038/41944.

[6] Dułak D, Gadzała M, Banach M, Ptak M, Wiśniowski Z, Konieczny L, Roterman I. Filamentous aggregates of Tau proteins fulfil standard amyloid criteria provided by the fuzzy oil drop (FOD) model. International Journal of Molecular Sciences 2018; 19(10):2910. https://doi.org/10.3390/ijms19102910.

CHAPTER 9

Non-amyloid structure of the Aβ(1−42) polypeptide in presence of a permanent chaperone

Mateusz Banach, Irena Roterman
Department of Bioinformatics and Telemedicine, Jagiellonian University—Medical College, Krakow, Poland

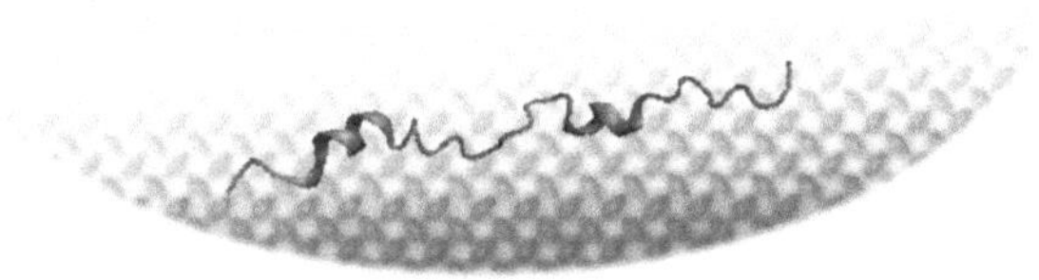

Conceptual diagram which illustrates structural stabilization of a polypeptide in the presence of an external molecule. The biologically active form requires interaction with a scaffold — a cellular membrane, another protein or a suitable compound.

The structure of the Aβ(1−42) amyloid has been compared with its non-amyloid form, with particular focus on fragments identified as amyloid seeds. The presented results enable us to identify conformational changes involved in the amyloid transformation process.

Ever since the discovery of prions, the structure of amyloid aggregations and the mechanisms by which they emerge have puzzled molecular biologists [1,2].

Radical conformational rearrangements in the absence of mutations appear to contradict the prevalent view that the protein's 3D structure is fully encoded in the composition of its chain [3].

Particular attention has been accorded to proteins in which conformational rearrangements produce structures capable of unrestricted growth, resulting in fibrillar complexes [4]. This phenomenon is of great importance in biological research since amyloid fibrils cause a variety of pathological

From Globular Proteins to Amyloids
ISBN: 978-0-08-102981-7
https://doi.org/10.1016/B978-0-08-102981-7.00012-9

conditions in living organisms [5]. In particular, they may accumulate in brain tissue, leading to neurodegeneration [6].

The insolubility of amyloids greatly hampers analysis of their structural properties. In order to overcome these obstacles prior to invention of solid-state NMR [7] amyloidogenic peptides were complexed with other molecules, including other proteins, in order to ensure solubility [8–11].

CHAPTER 9.A

Complexes Aβ(1−42) polypeptide with non-protein molecules

Mateusz Banach[1], **Irena Roterman**[1]
[1]Department of Bioinformatics and Telemedicine, Jagiellonian University—Medical College, Krakow, Poland

Contents

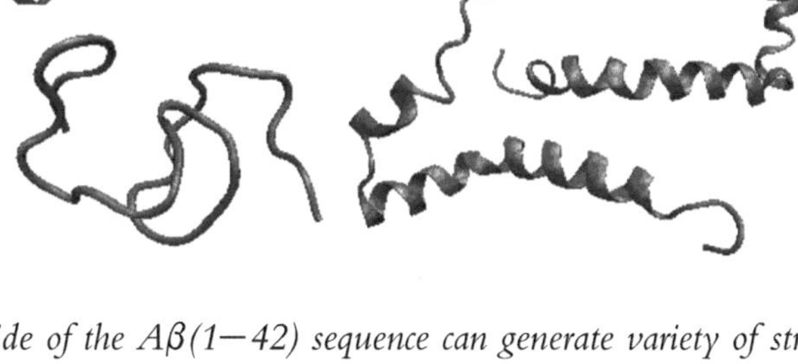

Polypeptide of the Aβ(1−42) sequence can generate variety of structures conditioned by environment participation.

From Globular Proteins to Amyloids
ISBN: 978-0-08-102981-7
https://doi.org/10.1016/B978-0-08-102981-7.00013-0

The presented analysis focuses on a selection of polypeptides all of which are fragments of Aβ(1–42). Their structure is determined in conditions which approximate the natural environment, using NMR techniques. Further analysis focuses on the distribution of hydrophobicity in native proteins, calculated using the fuzzy oil drop model [12]. The hydrophobicity profiles of various non-amyloid forms of Aβ(1–42) are compared with their amyloid counterparts [13]. The goal of this process is to identify common factors which may explain the underlying mechanism behind radical structural changes associated with amyloid transformation.

Structures discussed in this chapter differ significantly from the amyloid forms of Aβ(1–42). We consider proteins which do not produce long fibrils. These structures tend to contain helical folds which are not present in amyloids; additionally, their conformations reveal significant involvement of random coil fragments. Particular attention is devoted to fragments suspected of acting as amyloid seeds. In all subsequent descriptions the residue numbering pattern is consistent with Aβ(1–42).

Table 9.A.1 lists proteins subjected to analysis. All of these structures underwent NMR imaging in a membrane-mimicking environment (mostly SDS micelles) (SDS - Sodium dodecyl sulfate). The table includes proteins derived from *Homo sapiens* and *Rattus norvegicus*, as well as several synthetic constructs. Table 9.A.1 also lists conditions for structures which require external factors to ensure stabilization including DMSO (dimethyl sulfoxide).

When heated in the presence of SDS proteins unfold and, together with SDS molecules, reassemble into a spiral rod covered by SDS molecules (12 carbons and a sulfo group).

While DMSO denatures proteins, it does not directly act upon their structure – instead, the solvent acts upon water, binding its molecules in an exothermic process. Urea is a denaturing factor. When purged (via dialysis), the dissolved protein reverts to its native form. Ammonium sulfate promotes crystallization of proteins.

Structural characteristics of Aβ(1–42) fragments in complex with selected compounds

All proteins which comprise the study set represent non-amyloid forms of the Aβ(1–42) polypeptide, and all of them include helical fragments. This is schematically depicted in Fig. 9.A.1.

Table 9.A.1 List of proteins subjected to analysis in this chapter, along with their relation to the Aβ(1–42) sequence. The table also lists experimental conditions which enable stabilization of non-aggregating forms (including references).* – circular micelle formed on the surface of a drop of detergent.

PDB ID	FR.	Helix	External factors	Source	Ref.
1AMB	1–28	1–28	Membrane-like	*Homo sapiens*	[14]
1AMC	1–28	1–28	Membrane-like	*Homo sapiens*	[14]
1AML	1–40	14–24, 33–35	Trifluoromethanol	*Homo sapiens*	[15]
1BA4	1–40	15–36	SDS	*Homo sapiens*	[16]
1BA6	1–40	16–26	SDS	*Homo sapiens*	[17]
1BJB	1–28	15–25	SDS Mutation K16E	*Homo sapiens*	[18]
1BJC	1–28	16–25	SDS Mutation K16F	*Homo sapiens*	[18]
1HZ3	10–35	Random Coil	TSP trimethylsilylpropionate	Synthetic	[19]
1IYT	1–42	7–26, 27–40	Fluorinated alcohols	Synthetic	[20]
1NMJ	1–28	15–25	DMSO + Zn(2+)	Ratus norv.	[21]
1Z0Q	1–42	10–22, 28–32	Hexafluoroisopropanol	Synthetic	[22]
2LFM	1–40	12–18, 19–23	50mM NaCl	Synthetic	[23]
2LP1*	12–40	18–26, 27–42	LMPG(lyso-myristoylphosphatidylcholine)	*Homo sapiens*	[24]
2MJ1	17–34	19–24, 26–34	Glutamate – N-terminal Arginine – C-terminal	Synthetic	[25]

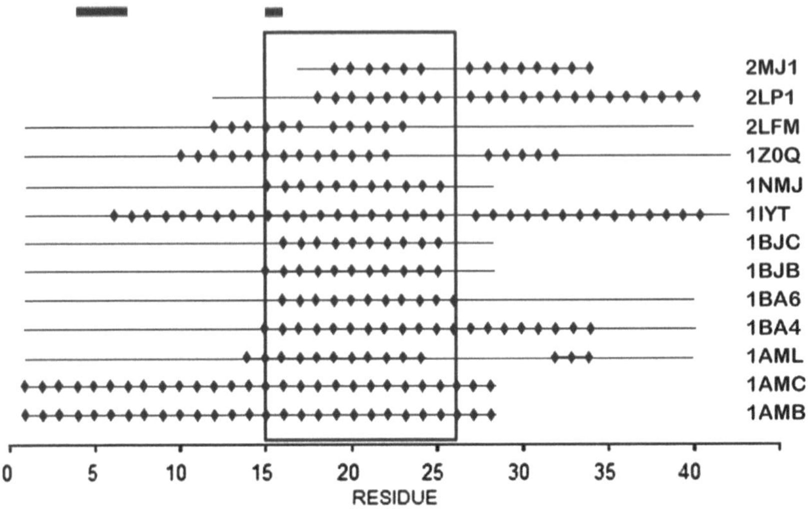

Fig. 9.A.1 Comparison of helical folds (diamonds) in the presented proteins. Blue frame highlights fragment 15–26 of Aβ(1–42) in each structure. Thick lines on top reveal the positions of tetrapeptides which (as suggested by analysis of chameleon sequences) are particularly prone to producing helical folds.

As shown in Fig. 9.A.1, all analyzed proteins contain helical structures within their 15–26 fragment. However, this fragment is not identified as particularly susceptible to producing a helical fold (fragments which fulfill this criterion are marked on the top of the figure, in accordance with a database of chameleon sequences [26]).

Structural analysis involving a set of amyloid structures produced by various fragments of Aβ(1–42) suggests that the 16–22 fragment has a highly peculiar structure and likely acts as an amyloid seed [13]. Other fragments which undergo conformational changes in amyloids are those at 11–16 and 22–28 [13]. Accordingly, our analysis will focus on these fragments.

Regarding the 16–22 fragment, it exhibits a helical conformation in all structures presented in this chapter, but does not retain this property in any amyloid form of Aβ(1–42) as it is shown in Chapter 10.A.

In order to simplify interpretation of the presented results we provide plots 9 Fig.9.A.2.) describing the distribution of hydrophobicity in three representative amyloids Aβ(15–40) (2MPZ [27], Aβ(1–40) (2MVX) [28] and Aβ(15–40) (2MXU) [29]. In all cases, the theoretical distribution (T) predicts two maxima separated by a hydrophilic linker. Comparing T and O reveals discordance in the area of residues 11–18 and 21–28; we will therefore attempt to determine why these fragments deviate from a

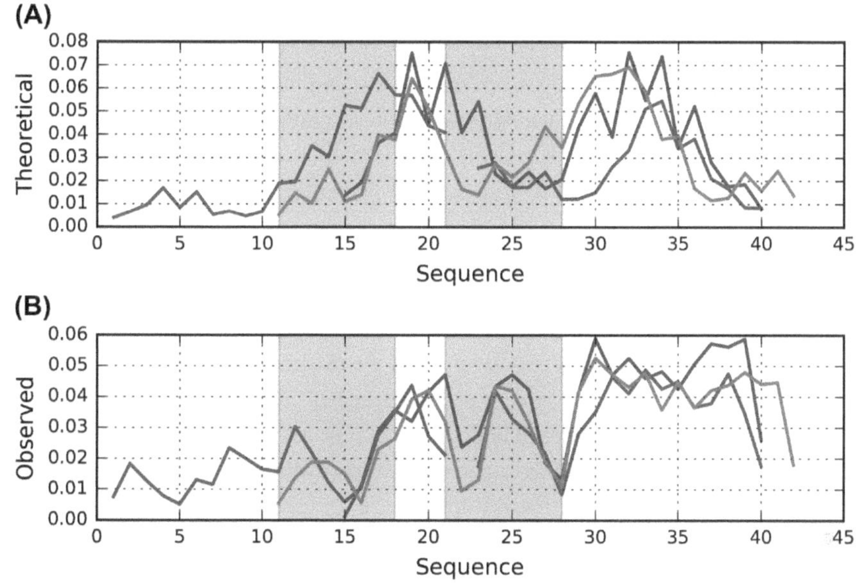

Fig. 9.A.2 Hydrophobicity distribution profiles for Aβ(15–40) – blue (2MPZ), Aβ(1–40) – red (2MVX) and Aβ(1–40) – green (2MXU) amyloids. Red background highlights two fragments (11–18 and 21–28) which exhibit discrepancies between hydrophobicity distributions. (A) theoretical distribution (T), (B) observed distribution (O).

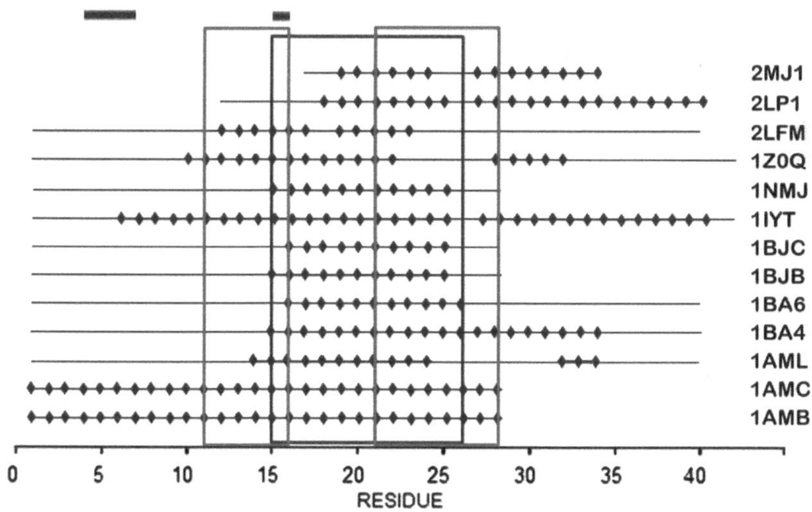

Fig. 9.A.3 Comparison of helical folds (diamonds) in the presented proteins. Blue frame highlights fragment 15–26 of Aβ(1–42) in each structure. Red frames highlight fragments 11–16 and 21–28 where O diverges from T in amyloid structures. Thick lines on top reveal the positions of tetrapeptides which (as suggested by analysis of chameleon sequences) are particularly prone to producing helical folds.

distribution which would otherwise ensure solubility and prevent formation of amyloid fibrils.

Fig. 9.A.3 reveals strong recurrence of helical folds (blue frame in Fig. 9.A.1). It turns out that amyloid seeds are located in closeness, proximity to such folds.

Assessment of the hydrophobic core structure in non-amyloid forms of Aβ(1–42)

Table 9.A.2 presents the outcome of our analysis, listing RD values (T-O-R and T-O-H variants) along with a summary of correlation coefficients corresponding to the presented fragments.

Two presented structures, 1HZ3 and 2LP1, are characterized as accordant with the theoretical distribution of hydrophobicity. This shows that fragments of the Aβ(1–42) polypeptide are indeed capable of producing

Table 9.A.2 Fuzzy oil drop parameters for the presented proteins. Values listed in boldface indicate structures which conform to the theoretical distribution of hydrophobicity. * – total chain length is 46 aa, including the 12–40 fragment of Aβ(1–42) at positions 1–29. The rightmost column lists residues for which HvT and TvO coefficients adopt negative values while the corresponding HvO coefficient remains high.

		RD		
PDB ID	**Chain**	**T-O-R**	**T-O-H**	**Negative**
1AMB	1–28	0.637	0.392	13–14
1AMC	1–28	0.597	0.372	4, 22
1AML	1–40	0.608	0.519	7–12, 22
1BA4	1–40	0.709	0.580	12, 23
1BA6	1–40	0.661	0.394	
1BJB	1–28	0.618	0.457	7–15
1BJC	1–28	0.610	0.540	8–10
1HZ3	10–35	**0.446**	0.312	13–19
1IYT	1–42	0.686	0.548	12–16, 26
1NMJ	1–28	0.576	0.399	9, 10, 20–22
1Z0Q	1–42	0.740	0.663	6–11, 18–29
2LFM	1–40	0.615	0.591	12–16, 18–23
2LP1	12–40*	**0.332**	0.309	2–5, 7–11
2MJ1	17–34	0.581	0.373	2–7

globular forms, with prominent concentrations of hydrophobicity at their cores, and solubility mediated by exposure of hydrophilic residues.

As already mentioned and further discussed in Ref. [13], the fragments at 11–16, 16–22 and 22–28 exhibit peculiar characteristics – different from what would be expected in a globular protein. Consequently, these fragments will be subjected to more detailed analysis.

Analysis of the 11–16 fragment

Regarding the fragment at 11–16, in its amyloid form the expected single peak of hydrophobicity (consistent with a globular structure) is replaced by two distinct peaks (Fig. 9.A.2). The Lys residue at position 16 causes strong reduction of hydrophobicity at the center of the expected hydrophobic core, giving rise to a peculiar hydrophobic "band", flanked by Glu11 and Lys16 (both of which are hydrophilic). Note that in an amyloid structure identical fragments tend to cluster together, which means that their local distributions of hydrophobicity propagate along the axis of the fibril. The status of the corresponding fragment in non-amyloid forms of Aβ(1–42) is illustrated in Table 9.A.3 and Fig. 9.A.4, the latter of which provides a visual representation of RD (both T-O-H and T-O-R – Fig. 9.A.4A) as well as FOD

Table 9.A.3 Fuzzy oil drop parameters for the 11–16 fragment of Aβ(1–42) in presented proteins. Values listed in boldface are consistent with an amyloid structure, while underlined values remain accordant with the Gaussian distribution of hydrophobicity.

PDB ID	RD		Correlation coefficient		
11–16	T-O-R	T-O-H	HvT	TvO	HvO
1AMB	0.413	0.079	0.527	0.506	0.925
1A.MC	0.418	0.057	0.659	0.502	0.941
1AML	**0.846**	0.210	**−0.250**	**−0.260**	**0.856**
1BA4	**0.748**	0.186	**−0.739**	**−0.388**	**0.650**
1BA6	**0.870**	0.150	**−0.547**	**−0.517**	**0.711**
1BJB	**0.788**	0.342	**−0.823**	**−0.317**	**0.702**
1BJC	**0.693**	**0.590**	**−0.157**	**0.099**	**0.784**
1HZ3	0.318	0.202	0.752	0.830	0.855
1IYT	0.498	0.073	−0.116	0.346	0.714
1NMJ	0.310	0.030	0.512	0.763	0.613
1Z0Q	0.608	0.127	−0.141	0.319	0.550
2LFM	**0.750**	0.344	**−0.501**	**−0.271**	**0.717**
2LP1	0.771	0.218	−0.950	−0.179	0.148
2MJ1	0.531	0.310	−0.168	0.172	0.172

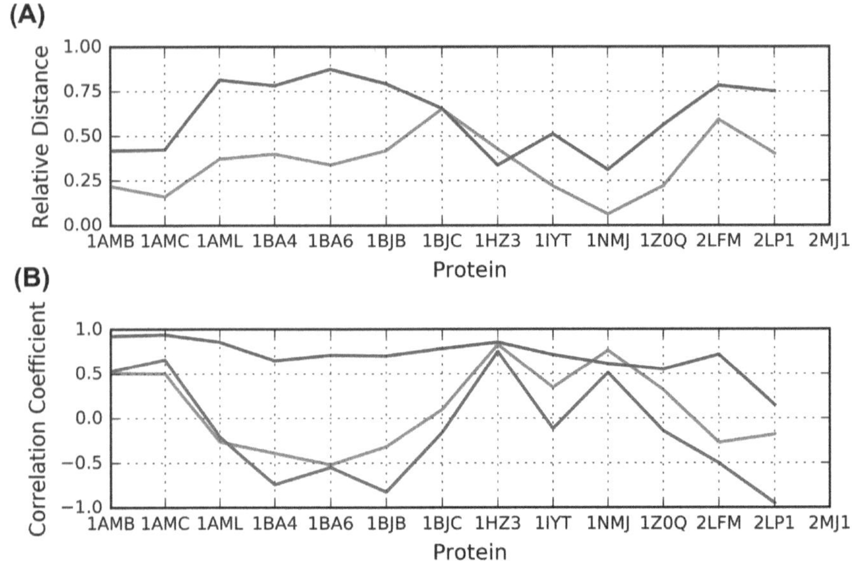

Fig. 9.A.4 FOD parameters calculated for the 11–16 fragment of Aβ(1–42) in presented proteins. (A) RD: T-O-R – blue, T-O-H – green. (B) correlation coefficients: HvT – red, TvO – green, HvO – blue.

correlation coefficients (Fig. 9.4.B) in each analyzed structure. In this context, we identify structures 1BA4, 1BA6 and 1BJB as particularly amyloid-like.

Analysis of the 16–22 fragment

This fragment of Aβ(1–42), when found in amyloid structures, exhibits a local peak of hydrophobicity corresponding to the second component of a monocentric hydrophobic core (expected in a globular protein). This local maximum, flanked by Lys16 on one side, and by Glu 22 and Asp 23 on the other, peaks at Phe19 and Phe20.

Table 9.A.4 presents the status of the 16–22 fragment in non-amyloid forms of Aβ(1–42), revealing structures which approach amyloid-like conditions. A graphical presentation is also provided in Fig. 9.A.3, showing values of RD (T-O-H and T-O-R – Fig. 9.A.3A) as well as FOD correlation coefficients (Fig. 9.A.3B).

Analysis of results presented in Table 9.A.4 and Fig. 9.A.5 singles out 1HZ3, 2LFM and 2LP1 as amyloid-like. However, in all presented proteins the 16–22 fragment remains helical. We may speculate that the specific distribution of hydrophobicity encoded by this fragment approximates

Table 9.A.4 Fuzzy oil drop parameters for the 16−22 fragment of Aβ(1−42) in presented proteins. Values listed in boldface are consistent with an amyloid structure, while underlined values remain accordant with the Gaussian distribution of hydrophobicity.

PDB ID	RD		Correlation coefficient		
16−22	**T-O-R**	**T-O-H**	**HvT**	**TvO**	**HvO**
1AMB	0.567	0.187	−0.079	0.038	0.584
1AMC	0.514	0.168	0.011	0.247	0.567
1AML	0.304	0.071	0.426	0.787	0.682
1BA4	0.538	0.204	0.032	0.087	-0.834
1BA6	0.631	0.185	0.065	−0.074	0.606
1BJB	0.625	0.366	0.164	−0.099	0.485
1BJC	0.630	0.328	−0.403	0.356	0.258
1HZ3	**0.666**	0.272	**−0.972**	**−0.547**	**0.637**
1IYT	0.546	0.055	0.138	0.381	0.645
1NMJ	0.425	0.157	0.460	0.589	0.711
1Z0Q	0.581	0.374	−0.139	0.127	0.919
2LFM	**0.662**	0.460	**−0.598**	**−0.447**	**0.934**
2LP1	**0.711**	0.201	**−0.411**	**−0.170**	**0.923**

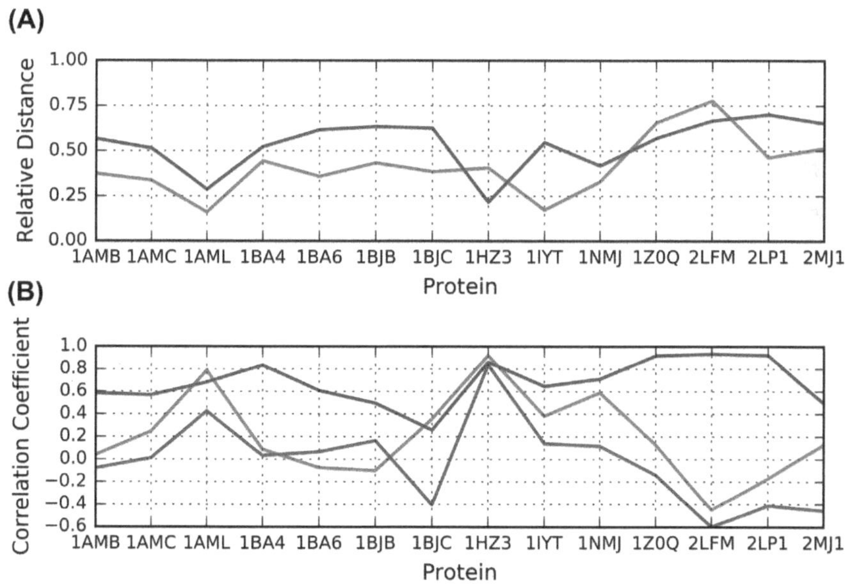

Fig. 9.A.5 FOD parameters calculated for the 16−22 fragment of Aβ(1−42) in presented proteins. (A) RD: T-O-R − blue, T-O-H − green. (B) correlation coefficients: HvT − red, TvO − green, HvO − blue.

conditions encountered in an amyloid fibril where no helical folds appear. Proteins listed in boldface in Table 9.A.4 are somewhat amyloid-like due to their high values of RD and disproportionate correlation coefficients, revealing strong involvement of intrinsic hydrophobicity. This is in contrast to proteins underlined in Table 9.A.4, whose parameters suggest micelle-like distribution.

Analysis of the 22–28 fragment

In amyloid structures this fragment contributes an additional local hydrophobicity maximum, which is particularly interesting given its location (in an area where the theoretical distribution expects hydrophobicity to remain low).

Table 9.A.5 illustrates the status of this fragment in native forms of Aβ(1–42), listing both RD values (T-O-R and T-O-H) along with all three correlation coefficients – HvT, TvO and HvO.

The corresponding visual depiction is provided in Fig. 9.A.6, which shows RD and correlation coefficients for each input structure. Notably, all examples listed in boldface in Table 9.A.5 exhibit amyloid-like conditions: high values of RD and diverging correlation coefficients, significantly biased toward HvO.

Table 9.A.5 Fuzzy oil drop parameters for the 22–28 fragment of Aβ(1–42) in presented proteins. Values listed in boldface are consistent with an amyloid structure, while underlined values remain accordant with the Gaussian distribution of hydrophobicity.

PDB ID	RD		Correlation coefficient		
22–28	T-O-R	T-O-H	HvT	TvO	HvO
1AMB	0.478	0.369	0.172	0.172	0.964
1AMC	<u>0.413</u>	<u>0.363</u>	<u>0.346</u>	<u>0.456</u>	<u>0.977</u>
1AML	**0.568**	0.309	**−0.281**	**0.018**	**0.944**
1BA4	<u>0.439</u>	<u>0.130</u>	<u>0.369</u>	<u>0.419</u>	<u>0.988</u>
1BA6	**0.747**	0.242	**−0.608**	**−0.435**	**0.937**
1BJB	0.468	0.489	0.110	0.279	0.944
1BJC	**0.668**	**0.588**	**−0.702**	**−0.419**	**0.862**
1HZ3	**0.604**	0.193	**−0.468**	**−0.359**	**0.883**
1IYT	0.394	0.126	0.527	−0.617	0.945
1NMJ	0.524	0.485	−0.024	0.295	0.913
1Z0Q	**0.642**	0.202	**−0.486**	**−0.534**	**0.990**
2LFM	0.553	0.426	−0.130	0.143	0.920
2LP1	**0.649**	**0.534**	**−0.131**	**−0.087**	**0.905**

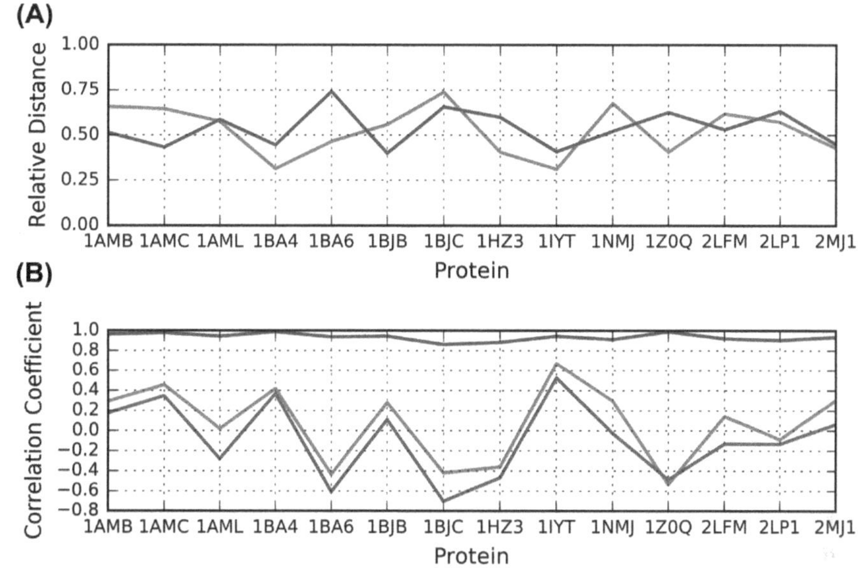

Fig. 9.A.6 FOD parameters calculated for the 22−28 fragment of Aβ(1−42) in presented proteins. (A) RD: T-O-R − blue, T-O-H − green. (B) correlation coefficients: HvT − red, TvO − green, HvO − blue.

Unusual structure observed in 1HZ3 − 10-35 fragment of Aβ(1−42) forming a random coil, together with an external factor (trimethylsilylpropionate; TSP)

The structure of 1HZ3 is particularly notable. This synthetic protein, created using a solid-phase method devised by Merrfield [30], exhibits a distribution of hydrophobicity which strongly corresponds to the theoretical (monocentric) Gaussian. The entirety of the chain is characterized as a random coil and

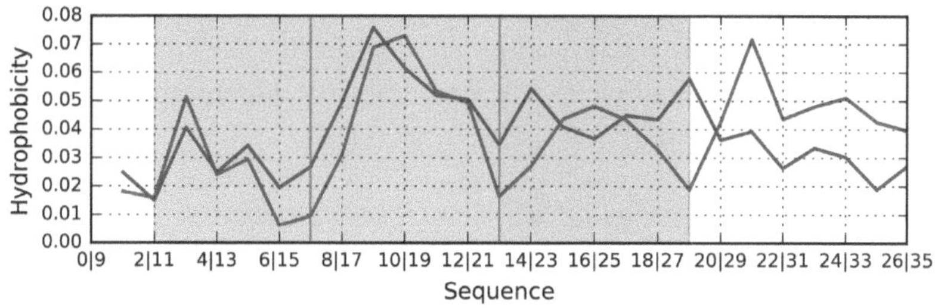

Fig. 9.A.7 Theoretical (T, blue) and observed (O, red) hydrophobicity distribution profiles for 1HZ3. Colored backgrounds correspond to fragments discussed in this chapter (blue − 11−16 and 16−22, red − 22−28). Dual labels on the horizontal axis list PDB residue numbers (first value) and positions relative to Aβ(1−42) (second value).

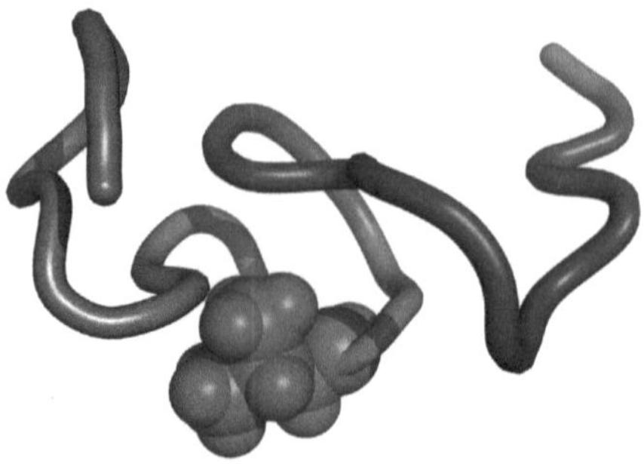

Fig. 9.A.8 3D presentation of 1HZE. Accordant fragments 11–16 and 16–22 are marked in blue while discordant fragment 22–28 is shown in red. Position of Val15 (Val24 in Aβ(1–42) amyloid) causing unexpected local hydrophobicity maximum in the 22–28 fragment is highlighted with space-filling display style.

packed into a globular structure which differs greatly from other members of the study set (all of which contain helical folds and are at least somewhat elongated). Fig. 9.A.7 provides a comparative overview of theoretical and observed distributions for this protein.

The fragments at 11–16 and 16–22 are both strongly accordant (which is also evident in Tables 9.A.2 and 9.A.3). The theoretical distribution predicts local peaks within each fragment, and such peaks are indeed observed in 1HZ3. On the other hand, despite the overall good alignment between T and O, the fragment at 22–28 reveals discordance which is characteristic of an amyloid seed, with negative values of HvT and TvO, and strongly positive values of HvO. In this fragment, the expected local minimum gives way to yet another local maximum. As a result, the entire structure – seemingly very distant from an amyloid (given its low values of RD) – is nevertheless susceptible to amyloid transformation.

Figs. 9.A.7 and 9.A.8 illustrate the location of the discordant fragment (red), with an exposed Val15 (in PDB sequence)/Val24 (in Aβ(1–42) amyloid) residue flanked by polar residues and locked in a very disadvantageous location (from the point of view of hydrophobicity distribution), producing an unexpected local peak.

Common characteristics of all discussed proteins

A characteristic property shared by almost all presented proteins is the presence of a local maximum corresponding to the 22–28 fragment (numbering is relative to Aβ(1–42), along with notable local maxima at 11–16 and 16–22 (Fig. 9.A.9). The chart reveals two outliers: 1AMC and 1AMB,

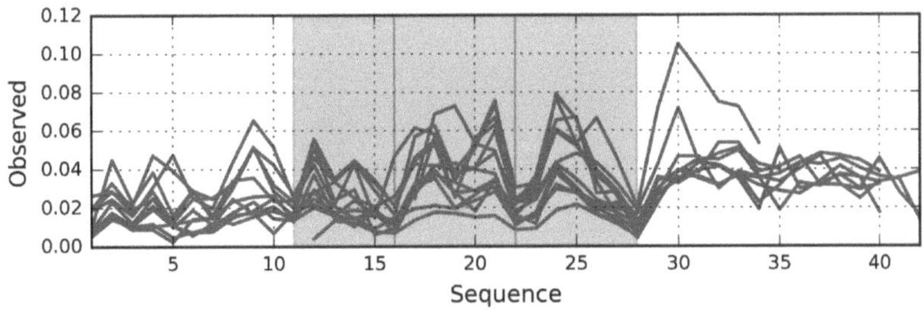

Fig. 9.A.9 Observed hydrophobicity distribution profiles for presented proteins. Red background highlights three fragments which are particularly useful in the context of Aβ(1−42) amyloid structure analysis (11−16, 16−22, 22−28).

Fig. 9.A.10 3D presentation of 1AMB (and likewise 1AMC) – an entirely helical structure.

where no local maximum can be observed at 22−28, while a single local maximum (instead of two) dominates the 11−22 fragment.

Given that the presented distributions characterize structures which differ in terms of tertiary conformations and chain lengths, their similarity should be regarded as notable.

The only two proteins which clearly deviate from the common pattern are 1AMB and 1AMC. This is due to their characteristic structure – both are helices, as illustrated in Fig. 9.A.10.

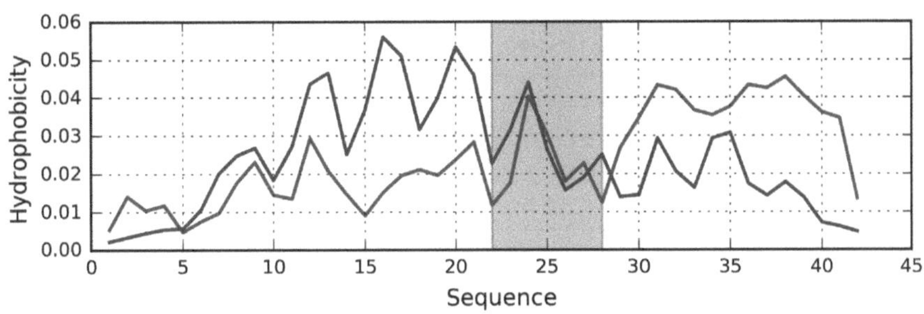

Fig. 9.A.11 Theoretical (T, blue) and observed (O, red) hydrophobicity distribution profiles for 1IYT, revealing local accordance between T and O in the 22−28 fragment.

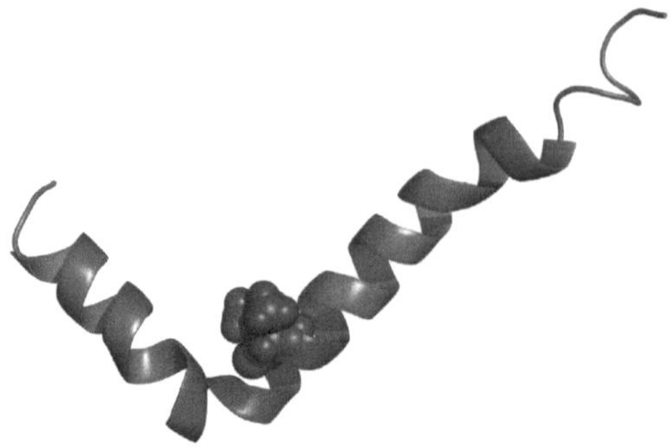

Fig. 9.A.12 3D presentation of 1IYT with 22–28 fragment marked in blue. While discordant fragment 22–28 is shown in red. Position of Val24 in the 22–28 fragment is highlighted with space-filling display style.

1IYT provides an example of a protein where a local maximum can be observed at 22–28. In this exceptional case the selected fragment (which is believed to act as an amyloid seed on the basis of other examples provided in this chapter) conforms to the theoretical distribution of hydrophobicity (Fig. 9.A.11). An explanation may be suggested by studying the corresponding 3D structure (Fig. 9.A.12).

The amphipathic nature of the helix is revealed in the 11–26 fragment where a local maximum is adjacent to a local minimum (Fig. 9.A.11). Fig. 9.A.12 highlights the location of the 22–28 fragment in the protein

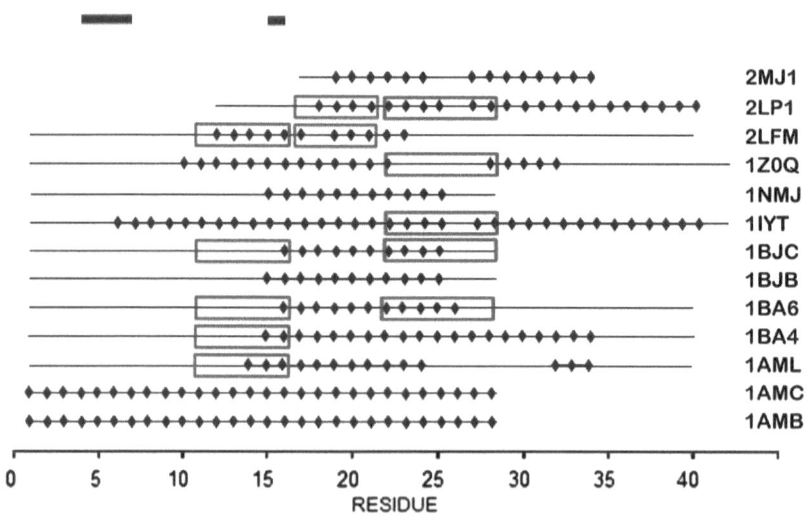

Fig. 9.A.13 Comparison of helical folds (diamonds) in the presented proteins. Blue frame highlights fragment 15–26 of Aβ(1–42) in each structure. Red frames – discordant fragments regarded as potential amyloid transformation seeds. Thick lines on top reveal the positions of tetrapeptides which (as suggested by analysis of chameleon sequences) are particularly prone to producing helical folds.

Table 9.A.6 Status of each analyzed fragment in presented proteins. X indicates conditions similar to those which are encountered in amyloids. The rightmost column lists the presence of external substances/entities.

	Fragments			
PDB ID	**11−16**	**16−22**	**22−28**	**External entity**
1AML	X			Membrane
1BA4	X			SDS
1BA6	X		X	SDS
1BJB	X			SDS
1BJC	X		X	SDS
1HZ3		X	X	TSP
1Z0Q			X	SDS
2LP1		X	X	SDS
2LFM	X	X		50mMNaCl

structure, which conforms to the theoretical distribution. The Val24 residue, facing the center of the encapsulating ellipsoid (3D Gaussian) is likewise accordant.

Comprehensive analysis

Fig. 9.A.13 and Table 9.A.6 describe the status of all three fragments (11−16, 16−22 and 22−28) in each of the analyzed proteins. This analysis shows the predisposition of certain short sequences to adopt structural forms observed in amyloids as well as in complex with external compounds.

Conclusions and discussion

In summary, it should be noted that the helical fragments found in the analyzed proteins do not undergo changes in terms of their contribution to the hydrophobic core structure (or lack thereof). In some sequences highly hydrophobic residues (Val12, Val18, Val24) are directly adjacent to polar residues (Gly11, Lys16, Glu22 and Lys28), resulting in a distribution of local maxima which does not conform to the theoretical Gaussian. Such drastic changes in hydrophobicity as between Lys/Glu and Val do not permit the chain to properly align itself with the Gaussian gradient. In theory, a hairpin structure with polar residues comprising the bend and hydrophobic residues making up either end should be oriented in such a way as to expose hydrophilicity on the surface while internalizing the hydrophobic terminal sections. In practice, however the Aβ(1−42) polypeptide is too short to generate a proper globular structure — such structures (i.e. in the case of type III antifreeze proteins) are more commonly identified for chains longer than 50 aa.

It is also interesting to note the structural patterns observed in 1BA4 and 1BA6. In both cases, at least two distinct fragments exhibit strongly amyloid-like properties. Similar amyloid seeds can be identified i.e. in 1BJB (11–16), 2LFM and 2LP1 (16–22) as well as 1BJC and 1Z0Q (22–28) by comparing them with amyloid forms of the Aβ(1–42) polypeptide.

Common phenomena observed in the presented structures include splitting the single local maximum into two less distinct maxima, along with the presence of an unexpected maximum roughly corresponding to the 22–28 fragment (even in 1HZ3, which otherwise shows excellent agreement between T and O).

Analysis of amyloid structures indicates the presence of local maxima in the 11–16, 16–22 and 22–28 fragments [13]. The cited work notes that such maxima are — in most cases — independent of the molecule's tertiary conformation, chain length or participation of helical folds. In some cases they approximate the theoretical distribution, resulting in a globular structure which is unable to form complexes under experimental conditions (i.e. in the presence of the SDS membrane-mimicking molecule). Nevertheless, when the environment promotes close contact between fragments which expose individual local maxima, this may be enough to kick-start the formation of a macromolecule consisting of many separate chains. Notably, all three local maxima discussed in this chapter are found in all amyloids listed in PDB, regardless of their chain length and secondary structural characteristics.

The goal of the above-mentioned work was to determine the presence of specific distributions of hydrophobicity in monomeric units which produce differing secondary folds. It turns out that the peculiar conditions found in amyloids are also replicated by monomeric proteins consistent with the Aβ(1–42) sequence.

It appears that — at least in some cases — preventing undesirable aggregation of monomers calls for external structures which act as chaperones, and that eliminating such structures usually results in the emergence of an amyloid. The list of external factors summarized in Ref. [16] clearly points to the critical role of the external environment in determining the final conformation of Aβ(1–42) fragments. This point is further addressed in Chapter 10. The fuzzy oil drop model enables us to provide a quantitative assessment of the impact of the native aqueous environment, i.e. the environment found in a properly functioning organism. This is done by reconciling the internal force field (represented by nonbinding interactions between atoms which make up the residue chain) and the external force field (contributed by the aqueous solvent, which promotes the formation of a centralized hydrophobic core). The identified need for various chaperones proves that Aβ(1–42) is not autonomous. In addition, the structure of an amyloid is found to oppose the influence the aqueous environment since instead of a central core we are faced with a linear pattern of fibrillar structures, based on the intrinsic hydrophobicity of the participating residues.

In light of the above, the presence of additional external compounds, such as SDS, DMSO, or others (see Tables 9.A.1 and 9.A.6) should be regarded as critical. These compounds either directly affect the structure of the protein or act by altering the properties of the solvent, which has an indirect effect on the conformation of the polypeptide. As already noted, the structure of the environment may, in some cases, lead to amyloid aggregation. On the other hand, some proteins require a non-aqueous environment to properly perform their function. Such proteins tend to misfold when placed in pure water, which – to them – represents an "alien" environment.

The presented group of proteins is classified as requiring a permanent chaperone. We can assume that the presence of a chaperone is needed to prevent conformational changes which would otherwise result in a misfolded protein. Here, instead of applying the traditional definition of a chaperone, we instead introduce the concept of "permanent chaperones", indicating that the aqueous solvent does not always represent the ideal environment for a folding chain. A separate question is why certain polypeptides, when separated from their respective permanent chaperones, evade natural degradation mechanisms and instead produce amyloid clusters. This phenomenon may be explained as an effect of improper (unexpected) structural properties of the surrounding environment.

In conclusion, it should be underscored that certain protein structures are only stable in the presence of environmental factors which may be referred to as permanent chaperones. When the chaperone disappears, the dependent polypeptide yields to its innate tendency to adopt a conformation determined by intrinsic hydrophobicity, without regard for the structural influence of the solvent [31]. It also appears that such effects are promoted by drastic changes in hydrophobicity over short fragments of the polypeptide chain (in particular, direct adjacency of strongly hydrophobic and strongly hydrophilic residues), which prevent the chain from aligning itself with the Gaussian hydrophobicity gradient. This phenomenon also seems more evident in short peptides, which are unable to produce a proper globular fold due to insufficient chain length.

References

[1] Prusiner SB, Bolton DC, Groth DF, Bowman KA, Cochran SP, McKinley MP. Further purification and characterization of scrapie prions. Biochemistry 1982; 21(26):6942–50. https://doi.org/10.1021/bi00269a050.

[2] Prusiner SB. Prions. Scientific American 1984;251(4):50–9. https://doi.org/10.1038/scientificamerican1084-50.

[3] Anfinsen CB. Principles that govern the folding of protein chains. Science 1973; 181(4096):223–30. https://doi.org/10.1126/science.181.4096.223.

[4] Tycko R. Physical and structural basis for polymorphism in amyloid fibrils. Protein Science 2014;23(11):1528–39. https://doi.org/10.1002/pro.2544.

[5] Eisenberg D, Jucker M. The amyloid state of proteins in human diseases. Cell 2012; 148(6):1188–203. https://doi.org/10.1016/j.cell.2012.02.022.

[6] Ross CA, Poirier MA. Protein aggregation and neurodegenerative disease. Nature Medicine 2004;10(7):S10–7. https://doi.org/10.1038/nm1066.

[7] Tycko R. Solid-State NMR studies of amyloid fibril structure. Annual Review of Physical Chemistry 2011;62(1):279–99. https://doi.org/10.1146/annurev-physchem-032210-103539.

[8] Richard T, Papastamoulis Y, Waffo-Teguo P, Monti J-P. 3D NMR structure of a complex between the amyloid beta peptide (1–40) and the polyphenol ε-viniferin glucoside: implications in Alzheimer's disease. Biochimica et Biophysica Acta (BBA) – General Subjects 2013;1830(11):5068–74. https://doi.org/10.1016/j.bbagen.2013.06.031.

[9] Kreutzer AG, Hamza IL, Spencer RK, Nowick JS. X-ray crystallographic structures of a trimer, dodecamer, and annular pore formed by an Aβ17–36 β-hairpin. Journal of the American Chemical Society 2016;138(13):4634–42. https://doi.org/10.1021/jacs.6b01332.

[10] Barrett PJ, Song Y, Van Horn WD, Hustedt EJ, Schafer JM, Hadziselimovic A, Sanders CR. The amyloid precursor protein has a flexible transmembrane domain and binds cholesterol. Science 2012;336(6085):1168–71. https://doi.org/10.1126/science.1219988.

[11] Hoyer W, Grönwall C, Jonsson A, Ståhl S, Härd T. Stabilization of a β-hairpin in monomeric Alzheimer's amyloid-β peptide inhibits amyloid formation. Proceedings of the National Academy of Sciences 2008;105(13):5099–104. https://doi.org/10.1073/pnas.0711731105.

[12] Kalinowska B, Banach M, Konieczny L, Roterman I. Application of divergence entropy to characterize the structure of the hydrophobic core in DNA interacting proteins. Entropy 2015;17(3):1477–507. https://doi.org/10.3390/e17031477.

[13] Dułak D, Banach M, Gadzała M, Konieczny L, Roterman I. Structural analysis of the Aβ(15–40) amyloid fibril based on hydrophobicity distribution. Acta Biochimica Polonica 2018. https://doi.org/10.18388/abp.2018_2647.

[14] Talafous J, Marcinowski KJ, Klopman G, Zagorski MG. Solution structure of residues 1–28 of the amyloid beta-peptide. Biochemistry 1994;33(25):7788–96. https://doi.org/10.1021/bi00191a006.

[15] Sticht H, Bayer P, Willbold D, Dames S, Hilbich C, Beyreuther K, Rosch P. Structure of amyloid A4-(1–40)-peptide of alzheimer's disease. European Journal of Biochemistry 1995;233(1):293–8. https://doi.org/10.1111/j.1432-1033.1995.293_1.x.

[16] Coles M, Bicknell W, Watson AA, Fairlie DP, Craik DJ. Solution structure of amyloid β-peptide(1–40) in a Water–Micelle environment. Is the membrane-spanning

domain where we think it is? Biochemistry 1998;37(31):11064–77. https://doi.org/10.1021/bi972979f.
[17] Watson AA, Fairlie DP, Craik DJ. Solution structure of methionine-oxidized amyloid β-peptide (1–40). Does oxidation affect conformational switching? Biochemistry 1998;37(37):12700–6. https://doi.org/10.1021/bi9810757.
[18] Poulsen S-A, Watson AA, Fairlie DP, Craik DJ. Solution structures in aqueous SDS micelles of two amyloid β peptides of aβ(1–28) mutated at the α-secretase cleavage site (K16E, K16F). Journal of Structural Biology 2000;130(2–3):142–52. https://doi.org/10.1006/jsbi.2000.4267.
[19] Zhang S, Iwata K, Lachenmann MJ, Peng JW, Li S, Stimson ER, Lee JP. The alzheimer's peptide aβ adopts a collapsed coil structure in water. Journal of Structural Biology 2000;130(2–3):130–41. https://doi.org/10.1006/jsbi.2000.4288.
[20] Crescenzi O, Tomaselli S, Guerrini R, Salvadori S, D'Ursi AM, Temussi PA, Picone D. Solution structure of the Alzheimer amyloid β-peptide (1–42) in an apolar microenvironment. European Journal of Biochemistry 2002;269(22):5642–8. https://doi.org/10.1046/j.1432-1033.2002.03271.x.
[21] Huang J, Yao Y, Lin J, Ye Y-H, Sun W-Y, Tang W-X. The solution structure of rat Aβ-(1–28) and its interaction with zinc ion: insights into the scarcity of amyloid deposition in aged rat brain. JBIC Journal of Biological Inorganic Chemistry 2004;9(5): 627–35. https://doi.org/10.1007/s00775-004-0556-x.
[22] Tomaselli S, Esposito V, Vangone P, van Nuland NAJ, Bonvin AMJJ, Guerrini R, Picone D. The α-to-β conformational transition of alzheimer's aβ-(1-42) peptide in aqueous media is reversible: a step by step conformational analysis suggests the location of β conformation seeding. ChemBioChem 2006;7(2):257–67. https://doi.org/10.1002/cbic.200500223.
[23] Vivekanandan S, Brender JR, Lee SY, Ramamoorthy A. A partially folded structure of amyloid-beta(1–40) in an aqueous environment. Biochemical and Biophysical Research Communications 2011;411(2):312–6. https://doi.org/10.1016/j.bbrc.2011.06.133.
[24] Barrett PJ, Song Y, Van Horn WD, Hustedt EJ, Schafer JM, Hadziselimovic A, Sanders CR. The amyloid precursor protein has a flexible transmembrane domain and binds cholesterol. Science 2012;336(6085):1168–71. https://doi.org/10.1126/science.1219988.
[25] Fonar G, Samson AO. NMR structure of the water soluble Aβ17–34peptide. Bioscience Reports 2014;34(6):759–64. https://doi.org/10.1042/bsr20140094.
[26] Ghozlane A, Joseph AP, Bornot A, de Brevern AG. Analysis of protein chameleon sequence characteristics. Bioinformation 2009;3(9):367–9. https://doi.org/10.6026/97320630003367.
[27] Sgourakis NG, Yau WM, Qiang W. Modeling an in-register, parallel "Iowa" aβ fibril structure using solid-state NMR data from labeled samples with rosetta. Structure 2015;23(1):216–27. https://doi.org/10.1016/j.str.2014.10.022.
[28] Schütz AK, Vagt T, Huber M, Ovchinnikova OY, Cadalbert R, Wall J, Güntert P, Böckmann A, Glockshuber R, Meier BH. Atomic-resolution three-dimensional structure of amyloid β fibrils bearing the Osaka mutation. Angewandte Chemie International Edition in English 2015;54(1):331–5. https://doi.org/10.1002/anie.201408598.
[29] Xiao Y, Ma B, McElheny D, Parthasarathy S, Long F, Hoshi M, Nussinov R, Ishii Y. Aβ(1–42) fibril structure illuminates self-recognition and replication of amyloid in Alzheimer's disease. Nature Structural and Molecular Biology 2015;22(6):499–505. https://doi.org/10.1038/nsmb.2991.

[30] Merrifield RB. Automated synthesis of peptides. Science 1965;150(3693):178–85. https://doi.org/10.1126/science.150.3693.178.
[31] Roterman I, Banach M, Konieczny L. Application of the fuzzy oil drop model describes amyloid as a ribbonlike micelle. Entropy 2017;19(4):167. https://doi.org/10.3390/e19040167.

CHAPTER 9.B

Structure of selected fragments of Aβ(1–42) in complex with other proteins

Mateusz Banach[1], Irena Roterman[1]
[1]Department of Bioinformatics and Telemedicine, Jagiellonian University–Medical College, Krakow, Poland

Contents

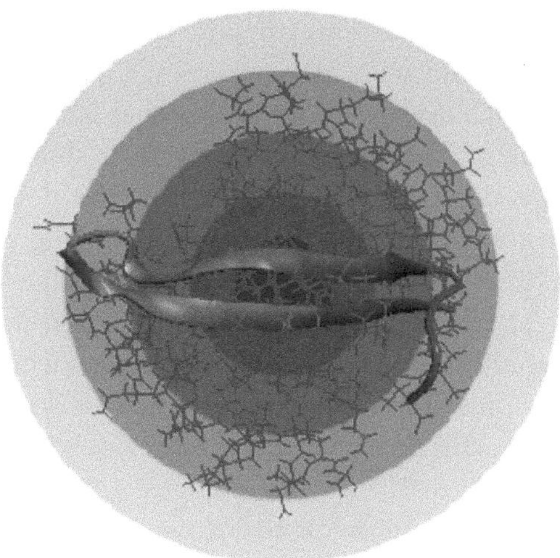

From Globular Proteins to Amyloids
ISBN: 978-0-08-102981-7
https://doi.org/10.1016/B978-0-08-102981-7.00014-2

Visualization of the stable structure of Aβ(16–40) conditioned by a permanent chaperone (external protein)

In our search for ways to prevent the generation of amyloid fibrils we have identified various chemical factors, including ligands and scaffold molecules, capable of binding individual fragments of the Aβ(1–42) polypeptide (see chapter 9A).

Techniques discussed in this chapter include complexation of external proteins [1,2], inserting fragments of amyloid proteins into other proteins – in particular the V domain of the IgG light chain [3] and complexation with ligands [4,5]. According to the authors' intentions, all these factors should result in the base polypeptide adopting conformations which do not produce amyloid fibrils. Analysis of such "arrested" peptides is based on the fuzzy oil drop model, where we compare the structure of the observed hydrophobic core with the corresponding theoretical distribution of hydrophobicity in an ideal protein.

The model is not always applicable – for instance, it fails to produce correct results for very short polypeptides which do not have a tertiary conformation. Nevertheless, it enables us to carry out comparative analysis by facilitating quantitative description of diverse structural forms. As a result, we may identify structural properties which promote amyloidogenesis.

Non-amyloid conformations of Aβ(1–42) fragments

This chapter discusses different fragments of the Aβ(1–42) peptide which – when complexed with other proteins – do not adopt amyloid-like conformations. Such external proteins (or other stabilizing factors) effectively act as "permanent chaperones", since in their absence the Aβ(1–42) chain may instead produce an amyloid-like form. The definition of a chaperone highlights its role in preventing the accompanying polypeptide from misfolding. Chaperones perform their function by temporarily disallowing specific structural rearrangements. The concept of a "permanent chaperone" builds upon this definition by removing temporal restrictions and assuming that the chaperone must be present at all times, as long as the protein performs its function. This property characterizes all structures discussed in this chapter.

The following proteins are discussed below:

1. Aβ(17–27) (PDB ID: 4MVL) – "packaged" with lipocain [1].
2. Aβ(16–28) (PDB ID: 4MVI) – in complex with lipocain [1].
3. Aβ(16–40) (2OTK) – "packaged" with a synthetic protein [2].

4. Aβ(16−28) (PDB ID: 2M9R and 2M9S) − in complex with the Y23 ligand *(2s,3s)-3-(3,5-Dihydroxyphenyl)-2-(4-Hydroxyphenyl)-4- [(E)-2-(4-Hydroxyphenyl)ethenyl]-2,3-Dihydro-1- Benzofuran-6-Yl beta-D-Glucopyranoside [Polyphenol epsilon-Viniferin glucoside]*, also referred to as polyphenol ε-viniferin glucoside (EVG) [3].
5. Aβ(1−21) (PDB ID: 5HOX) − in complex with the JEF ligand − *O-(O-(2-Aminopropyl)-O'-(2-Methoxyethyl)polypropylene glycol 500) [Jeffamine]* [4]
6. Aβ(18−41) (PDB ID: 3MOQ) − fragment incorporated into the IgG V domain (hypervariable loop at 88−111 − position in V domain chain) [5].

Aβ(17−27) and Aβ(16−28) in complex with lipocain

Lipocain, a versatile protein, has been used to bind fragments (17−27) and (16−28) of the Aβ(1−42) polypeptide, rendering it soluble. It is expected that such fragments may appear at early stages of degradation of large integral membrane proteins, potentially leading to formation of amyloid fibrils. β-secretase and γ-secretase activity results in production of lipophilic Aβ peptides with a total length of 42 or 40 residues (fragments at 672−711/713). The authors hypothesize that the resulting peptides may be "arrested" using lipocain.

Our analysis concerns two complexes, involving Aβ(17−27) and Aβ(18−28) respectively (Fig. 9.B.1). Chain A is contributed by lipocain, while chain B represents the target peptide (13 aa − 4MVI and 11 aa − 4MVL) [1]. Table 9.B.1 illustrates FOD parameters for the entire complex

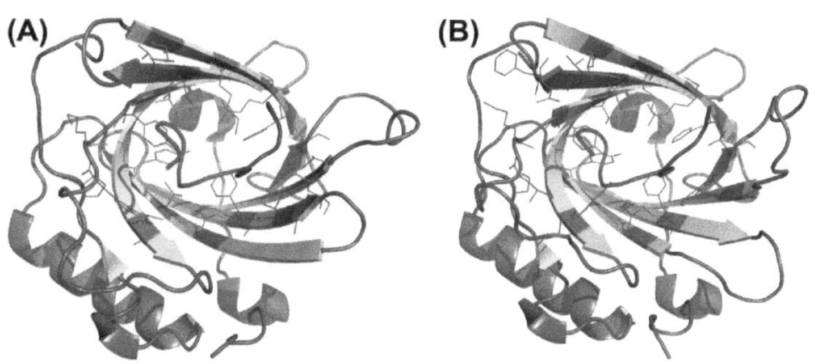

Fig.9.B.1 3D presentation of 4MVI and 4MLV complexes. (A) 4MVI chain A in complex with Aβ(16−28) peptide (chain B). (B) 4MVL chain A in complex with Aβ(17−27) peptide (chain E).Fragments 16−22 in Aβ(16−28) and 17−22 in Aβ(17−27) are shown in red. Fragments 22−28 in Aβ(16−28) and 22−27 in Aβ(17−27) are shown in blue. Residues in chain A in each protein engaged in protein-protein interactions with Aβ fragments are magenta-colored and have side chains displayed.

Table 9.B.1 Fuzzy oil drop parameters for 4MVI (with Aβ(16–28) peptide) and 4MVL (with Aβ(17–27) peptide. Asterisks mark the absence of Val18, causing a discordance. Values given in bold distinguish the status of discordance in respect to idealized hydrophobicity distribution.

		RD		Correlation coefficient		
PDB ID	**Fragment**	**T-O-R**	**T-O-H**	**HvT**	**TvO**	**HvO**
4MVI	Complex A + B	0.337	0.291	0.440	0.700	0.765
	Chain A	0.300	0.256	0.477	0.762	0.761
	Chain B	0.419	0.385	0.239	0.453	0.886
4MVI	Aβ(16–28)	0.372	0.256	0.617	0.667	0.873
Chain B	16-22	0.275	0.275	0.660	0.603	0.883
	22–28	0.227	0.227	0.869	0.886	0.933
4MVL	Complex A + E	0.319	0.290	0.449	0.709	0.747
	Chain E	0.482	0.400	−0.160	0.333	0.714
	Chain A	0.277	0.254	0.487	0.777	0.757
4MVL	Aβ (17–27)	0.446	0.400	0.173	0.513	0.705
Chain E	**17–22**	**0.595**	**0.405**	**−0.013**	**0.104**	**0.615**
	22–28	0.236	0.270	0.630	0.893	0.859
	22–28*	0.306	0.135	0.195	0.837	0.612

(lipocain + peptide), for the peptide analyzed as part of the complex, and for the standalone peptide.

As listed in Table 9.B.1, the complex as a whole exhibits low RD (Fig. 9.B.2). The Aβ chain, analyzed as a standalone unit (with a custom Gaussian capsule), is also accordant with the theoretical distribution (Fig. 9.B.3). Further analysis reveals, however, that the status of individual fragments varies. In the case of Aβ(17–27) the 17–22 fragment has amyloid-like properties: relatively high RD, low HvT and TvO, and high HvO.

Aβ(16–40) in complex with phage-display selected affibody protein Z(Aβ3)

We will now focus on the structure of Aβ(16–40) in complex with phage-display selected affibody protein Z(Aβ3) – engineered binding protein. The structure is listed in PDB as 2OTK [2].

Protein Z consists of two chains, 43 aa each, and acts as a packaging for the centrally placed Aβ(1–40) peptide. The peptide itself becomes a β-hairpin "arrested" between two mainly alpha up-down bundles (CATH code: 1.20.5.420), as defined in Ref. [6]. Both Z chains are listed as 14–56 fragments, linked with a disulfide bond (Cys28).

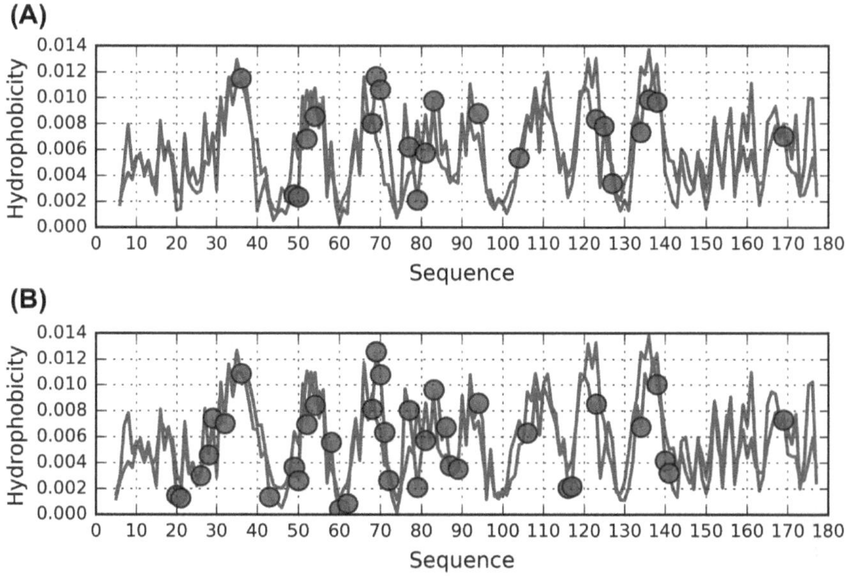

Fig. 9.B.2 Theoretical (T, blue) and observed (O, red) hydrophobicity distribution profiles for chains A of 4MVI and 4MVL analyzed as a part of complex with Aβ(1–42) peptides. (A) 4MVI. (B) 4MVL. Magenta circles denote residues engaged in protein-peptide interactions.

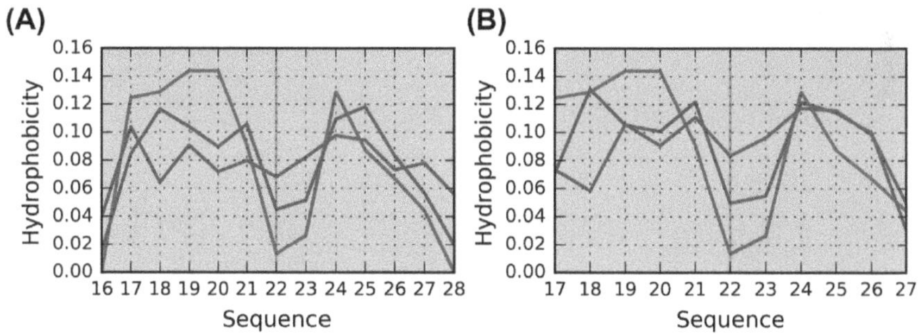

Fig. 9.B.3 Theoretical (T, blue), observed (O, red) and intrinsic (H, green) hydrophobicity distribution profiles for Aβ(1–42) peptides from 4MVI and 4MVL analyzed as individual units. (A) Aβ(16–28) – chain B (B) Aβ(17–27) – chain E. Colored background denotes fragments which correspond to the building blocks of Aβ(1–42) amyloid structures: 16/17–22 – red, 22–27/28 – blue.

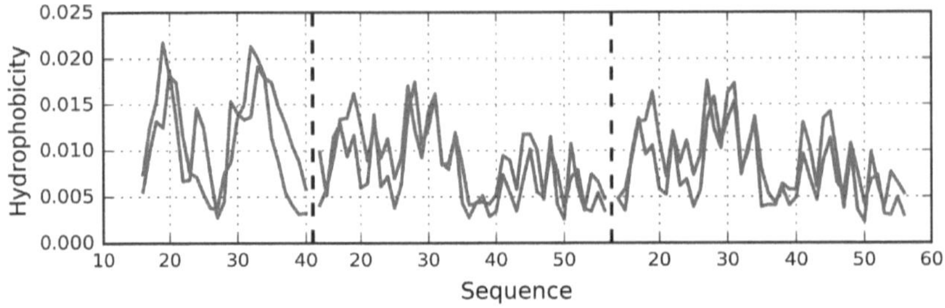

Fig. 9.B.4 Theoretical (T, blue) and observed (O, red) hydrophobicity distribution profiles for 2OTK complex. Dashed vertical lines mark boundaries between the chains (from left to right: C, E, F). Chain C represents the Aβ(16–40) peptide. Protein-protein contacts are not shown here due to the fact that nearly all residues from chain C interact with the other two chains.

Chains E and F belong to protein Z, while chain C is the Aβ(16–40) peptide.

Fig. 9.B.4 illustrates the theoretical and observed hydrophobicity distribution in the Z(Abeta3) complex. Note the similarities between protein Z chains (E and F) and the differing C chain.

Table 9.B.2 provides quantitative data – specifically, RD parameters and correlation coefficients for all presented structures.

Table 9.B.2 Fuzzy oil drop parameters for 2OTK. Chain C corresponds to Aβ(16–40). Values listed in boldface mark deviations from theoretical predictions in favor of amyloid-like conditions.

		RD		Correlation coefficient		
PDB ID	**Fragment**	**T-O-R**	**T-O-H**	**HvT**	**TvO**	**HvO**
2OTK	Complex E + F + C	0.407	0.328	0.558	0.702	0.803
	Complex E + F	0.370	0.335	0.622	0.728	0.864
	Chain E	0.391	0.340	0.603	0.880	0.709
Chain C	In complex	0.660	0.319	0.397	0.613	0.650
	16-22	0.241	0.049	0.804	0.776	0.824
	22–28	**0.578**	0.262	**−0.394**	**0.055**	**0.836**
	28–40	0.700	0.572	0.171	0.720	-0.173
	28–32	0.522	0.163	0.808	0.456	0.816
	32–40	0.675	0.583	0.140	0.764	-0.130
Chain C	Individual	0.591	0.290	0.279	0.493	0.602
	16–22	0.694	0.177	0.164	0.346	0.884
	22–28	**0.613**	0.356	**−0.305**	**−0.018**	**0.866**
	28–40	0.571	0.237	0.220	0.658	0.260

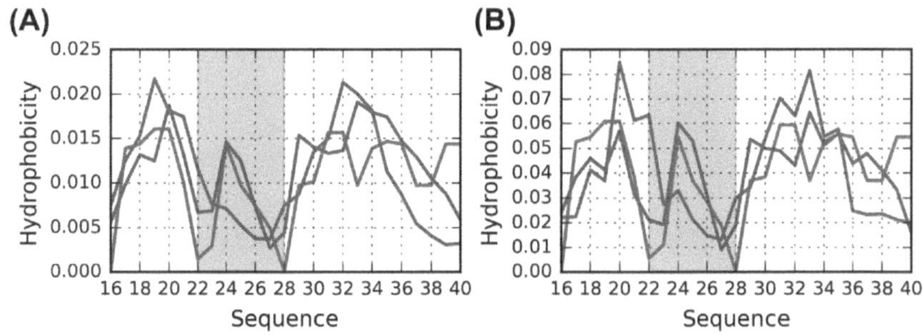

Fig. 9.B.5 Theoretical (T, blue), observed (O, red) and intrinsic (H, green) hydrophobicity distribution profiles for chain C of 2OTK corresponding to Aβ(16−40). (A) as a component of the Z complex. (B) as a standalone unit. The 22-28 fragment is highlighted as a potential amyloid seed due to its strong discordance.

Values listed in Table 9.B.2 suggest that the complex as a whole contains an ordered hydrophobic core, as evidenced by low RD and balanced correlation coefficients. The same is true for chains E and F, which "package" chain C.

Regarding chain C itself (Fig. 9.B.5), it deviates from the theoretical distribution both as a component of the protein complex and as a standalone structure. In attempting to identify the causes of this discordance we have computed RD parameters and correlation coefficients for fragments of the Aβ(16−40) chain which meet the amyloid seed criteria. These values, listed in Table 9.B.2, indicate that the fragment at 23−28 may be characterized as amyloid-like, with high RD(T-O-R), high HvO and very low HvT and TvO, even dipping into the negative territory. All these properties are consistent with an amyloid seed [7,8].

While the C chain as a whole is aligned with the structure of the Z complex, the 22−28 fragment retains its peculiar amyloid-like properties, similar to those observed in the Aβ(16−40) amyloid. Its exposed location within the complex is highlighted in Fig. 9.B.6.

Aβ(1−40) in complex with polyphenol ε-viniferin glucoside (EVG)

PDB lists two forms of Aβ(1−40) in complex with polyphenol ε-viniferin glucoside (EVG), differing with respect to the interaction site (i.e. the residues which engage the ligand): 2M9R and 2M9S [3].

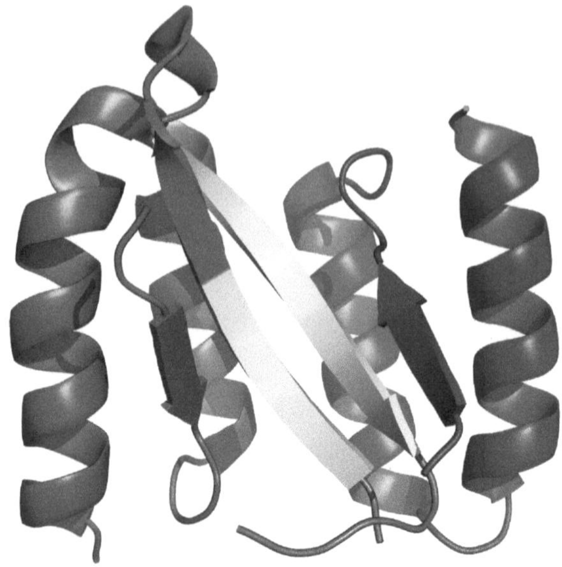

Fig. 9.B.6 3D presentation of 2OTK complex of three chains: E (blue), F (marine) and C – the Aβ(16–40) peptide (gray/yellow). Red color marks the 22–28 fragment of this chain. Protein-protein contacts are not shown here due to the fact that nearly all residues from chain C interact with the other two chains.

Table 9.B.3 provides a quantitative summary of both forms. In each case RD remains very high as a result of the random-coil conformation adopted by the chain. Interpretation of the remaining parameters is essentially out of scope of the fuzzy oil drop model, since no tertiary structure can be observed. As already suggested, the model is invoked here only to provide a common platform for comparative structural analysis.

Table 9.B.3 Fuzzy oil drop parameters for 2M9R and 2M9S. The residue numbers given in bold distinguish those of amyloid character in amyloid fibrils (Chapter 10).

		RD		**Correlation coefficient**		
PDB ID	**Fragment**	**T-O-R**	**T-O-H**	**HvT**	**TvO**	**HvO**
2M9R		0.852	0.857	0.050	−0.051	0.630
	16–22	0.447	0.284	0.463	0.410	0.895
	22-28	0.490	0.146	0.490	0.405	0.762
	29-40	0.968	0.913	0.130	−0.046	−0.398
2M9S		0.789	0.769	−0.022	−0.071	−0.632
	16-22	0.455	0.544	0.545	0.537	0.952
	22-28	0.366	0.218	0.521	0.717	0.835
	29-40	0.966	0.884	0.080	−0.406	−0.391

It is difficult to directly interpret the values listed in Table 9.B.3 due to the lack of a tertiary conformation; nevertheless, we may conclude that this protein does not appear to contain any hydrophobic core, and that it also does not produce an amyloid-like structure.

While the theoretical distribution of hydrophobicity is expected to contain one global maximum, multiple local maxima are observed instead (Fig. 9.B.7). This replacement of a large maximum with several smaller ones is a characteristic property of the Aβ(1−40) polypeptide, and is caused by the Lys residue at position 22, where high hydrophobicity is expected. A local maximum exists between residues 22 and 28. Fragments listed in boldface in Table 9.B.3 are highly discordant in amyloid conformations of Aβ(1−40) − however, they do not reveal strong discordance in either 2M9R or 2M9S.

Ligands play a crucial role in the analyzed polypeptides. The EVG molecule is strongly hydrophobic, and may act as a "substitute" hydrophobic core for the polypeptide chain. Such quasi-cores, contributed by ligands (rather than by the aqueous environment), may have a profound impact on the conformations of polypeptide chains, as evidenced by RD values for both ligand-binding and non-ligand-binding residues. As a result of the presence of EVG, the molecule becomes soluble. This suggests a possible design strategy for anti-fibril drugs.

The presence of relatively high hydrophobicity in the unstructured (loose) C-terminal fragment is intriguing.

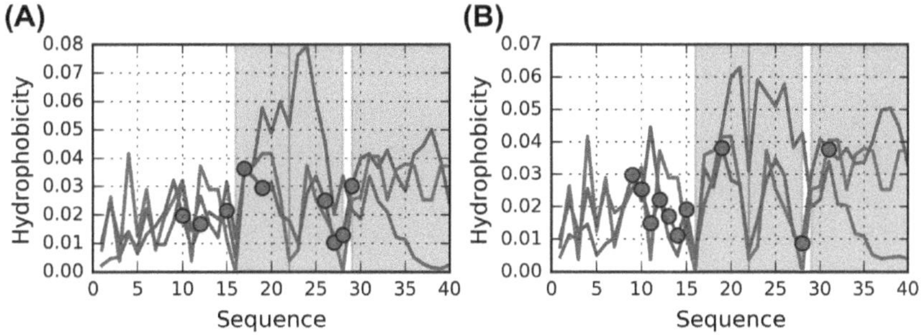

Fig. 9.B.7 Theoretical (T, blue), observed (O, red) and intrinsic (H, green) hydrophobicity distribution profiles for Aβ(1−40) complexes 2M9R and 2M9S. (A) 2M9R. (B) 2M9S. Blue background marks fragments 16-22 and 22-28 which are identified as amyloid seed in amyloid fibrils generated by Aβ(1-42) amyloid (Chapter 10A). Green circles denote resodies engaged in protein-ligand interaction.

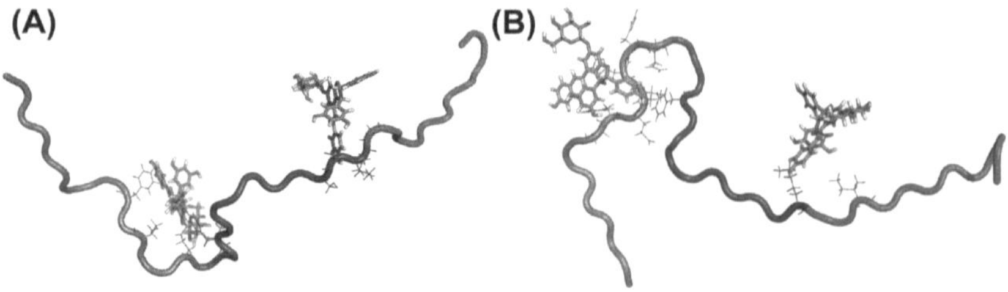

Fig. 9.B.8 3D presentation of 2M9R and 2M9S Aβ(1−40) complexes with EVG ligand. (A) 2M9R. (B) 2M9S. Fragments 16-22 and 22-28 which are treated as amyloid seeds (Chapter 10A) are given in red Residues engaged in ligand complexation are given in form of side-chains displayed.

The role of polyphenol ε-viniferin glucoside (EVG) as a permanent chaperone involves interaction with the central part of the chain (Fig. 9.B.8), producing a conformation which prevents the protein from aggregating into long amyloid-like fibrils.

In summary, it should be underscored that the presence of ligand is essential in preventing certain structural forms of Aβ(1−40) from amyloid transformation. The Lys residue at position 22 introduces structural disorder, which makes it difficult to generate a centralized hydrophobic core. This explains the presence of two local hydrophobicity maxima instead of a single global maximum. Engagement of fragments located on either site of the Lys residue prevents linear propagation, which would otherwise be possible. Thus, the ligand exerts a profound influence on the conformation of the presented chains. In line with the hypothesis formulated in Ref. [3], EVG appears to play the role of a protector against amyloid formation.

The presented analysis may be criticized as inadequate given that the sample polypeptides are incapable of producing globular structures (which is a prerequisite of applicability of the fuzzy oil drop model). Applicability of fuzzy oil drop model to molecules which do not generate the tertiary structure is not reasonable. This analysis was performed anyway to make the analysis of polypeptides discussed in this chapter complete. This example can be taken as the example to complete wide spectrum of different structural forms treated as targets for fuzzy oil drop model. However, we have decided to include them in order to broaden our study of the various forms adopted by the Aβ(1−40) polypeptide. In particular, comparing amyloid-like and non-amyloid-like structures which share identical sequences is of great theoretical interest. The applied model enables assessment of arbitrary molecules with regard to their hydrophobicity distribution − a factor strongly implicated in amyloidogenesis [7,8].

Aβ(1−21) in complex with the JEF ligand

The PDB file with ID 5HOX lists the Aβ(1−21) chain in a hexamer (six β-hairpins) complex with the JEF ligand − *O-(O-(2-Aminopropyl)-O'-(2-Methoxyethyl)polypropylene glycol 500) [Jeffamine]* ($C_{30}H_{63}NO_{10}$) [4]. All hairpins have the same sequence and share a similar structure. The ligand is centrally located, which suggests its role in coordinating the complex.

Table 9.B.4, (Figs. 9.B.9 and 9.B.10) characterizes the whole peptide as well as selected structural folds comprising the β-hairpin. The fragment at 2−10 exhibits strong amyloid-like properties (Fig. 9.B.9). To enable

Table 9.B.4 Fuzzy oil drop parameters for chain A of 2OTK. Values listed in boldface deviate from theoretical predictions in favor of amyloid-like conditions. The underlined fragment is regarded as weakly amyloid-like.

		RD		Correlation coefficient		
PDB ID	Fragment	T-O-R	T-O-H	HvT	TvO	HvO
5HOX	1-21	0.554	0.216	0.204	0.603	0.767
	2-10	**0.803**	**0.545**	**−0.455**	**−0.252**	**0.762**
	13-21	0.484	0.116	0.221	0.796	0.657
	(2−10) + (13−21)	**0.677**	**0.282**	**−0.092**	**0.390**	**0.697**
	16-21	0.655	0.580	−0.167	0.896	0.131

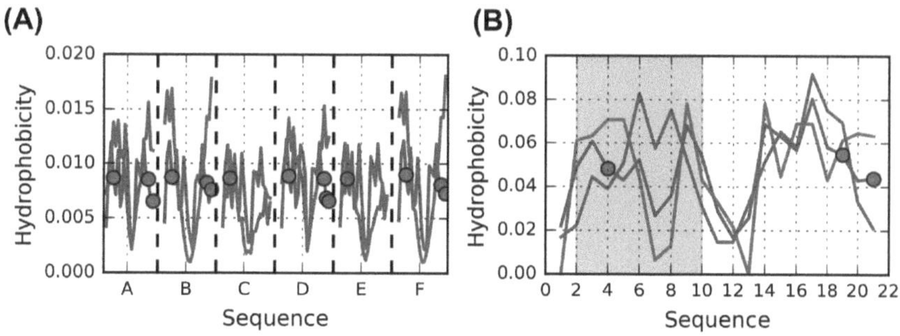

Fig. 9.B.9 Theoretical (T, blue), observed (O, red) and intrinsic (H, green) hydrophobicity distribution profiles for 5HOX. (A) for every chain as a part of the complex. (B) for chain A as individual unit. Red background distinguishes fragment at 2−10 with strong amyloid-like properties. Green circles denote residues engaged in protein-ligand interactions.

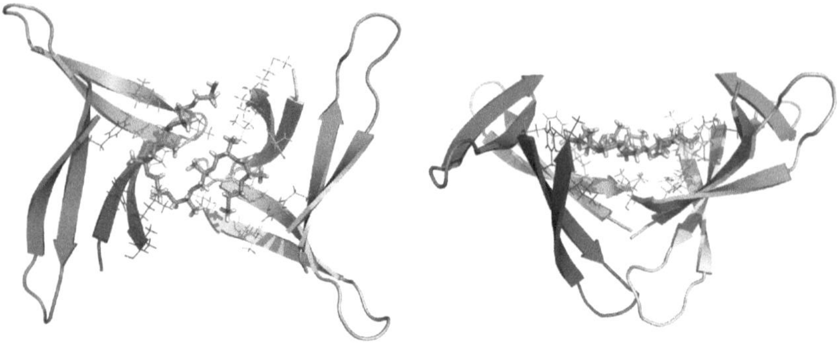

Fig. 9.B.10 3D presentation of 5HOX hexamer from two angles, revealing the crucial role of the JEF ligand in coordinating the β-hairpin complex. Each chain is presented in different color. Residues engaged in protein-ligand interactions have side chains displayed.

comparisons with other peptides and amyloids, we also distinguish the fragment at 16–21, which may be described as weakly amyloid-like.

The deviation identified at 2–10 consists of two local maxima where only a single maximum is predicted by the theoretical model. The existing local maxima are also broader than the expected single maximum. For example, at position 9 the theoretical distribution expects a gradual decline in hydrophobicity; however a local peak is observed instead – effectively "in opposition" to T. Such conditions are common in amyloids.

Aβ(18–41) fragment inserted as the antigen receptor variable domain in shark immunoglobulin

In this example, the Aβ(18–41) fragment constitutes the antigen receptor variable domain in shark immunoglobulin [5]. It is listed in PDB as 3MOQ.

From among all fragments discussed in this chapter this particular fragment has the fewest degrees of freedom. It is inserted into the V domain of immunoglobulin G, which fixes its N- and C-terminal sections – even though its overall length (33 aa) admits flexible rearrangements, since the fragment is classified as a hypervariable loop, which must align itself to the target antigen in order to perform its biological function.

As expected by the authors, a non-amyloid structural form of Aβ(18–41) has been obtained and found to resist aggregation. While fixed on both ends, it nevertheless retains some structural flexibility, enabling us to study its conformational preferences (Table 9.B.5).

Table 9.B.5 Fuzzy oil drop parameters for 5HOX. Underlined values indicate a weakly amyloid-like fragment.

		RD		Correlation coefficient		
PDB ID	**Fragment**	**T-O-R**	**T-O-H**	**HvT**	**TvO**	**HvO**
3MOQ	88−111 Aβ(18−41)	0.682	0.691	0.093	0.302	0.670
	18−22	0.207	0.496	0.680	0.830	0.818
	22−28	0.511	0.299	−0.105	0.340	0.781

The characteristic presence of Lys98 seen in Fig. 9.B.11 creates a local minimum where a participation in maximum is expected instead (92−106). However in this structure, the lysine residue is only partially exposed on the surface (Fig. 9.B.12) remaining close to the center, where hydrophobicity should otherwise remain high. This amyloid-like deviation, while not clearly revealed in the whole-domain plot, becomes notable when the Aβ(18−41) fragment is studied on its own.

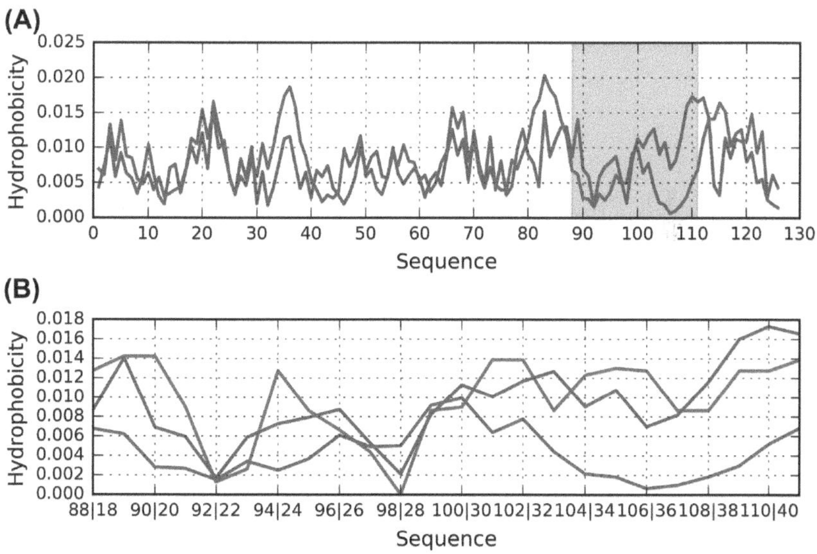

Fig. 9.B.11 Theoretical (T, blue), observed (O, red) and intrinsic (H, green) hydrophobicity distribution profiles for 3MOQ. (A) complete chain with Aβ(18−41) fragment highlighted in red. (B) Aβ(18−41) fragment (residues 88−111). Dual labels on the horizontal axis of B list PDB residue numbers (first value) and positions relative to Aβ(18-41) (second value).

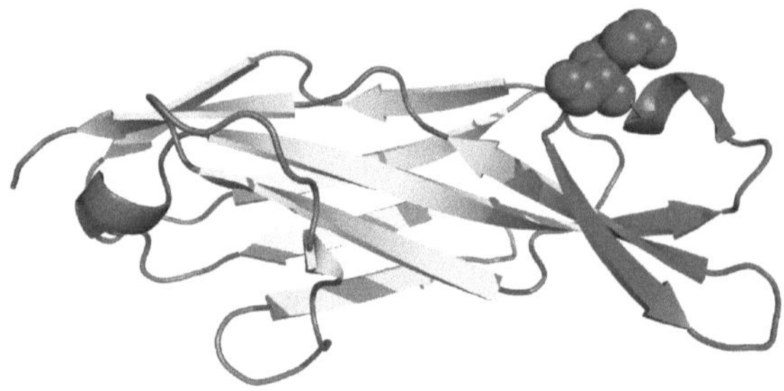

Fig. 9.B.12 3D presentation of 3MOQ, with the Aβ(18–41) fragment (residues 88–111) marked in red. Position of Lys98 in this fragment is highlighted with space-filling display style.

Discussion and conclusions

Fig. 9.B.13 provides a comparative overview of the similarities and differences between proteins discussed in this chapter.

Fig. 9.B.13 underscores the key role of the ligand or peptide with which the protein interacts. β-fragments found in 2OTK and 5HOX are almost fully engaged in interaction with the encapsulating protein – their permanent chaperone. 4MVI, 4MVL, 2M9R and 2M9S are all characterized as

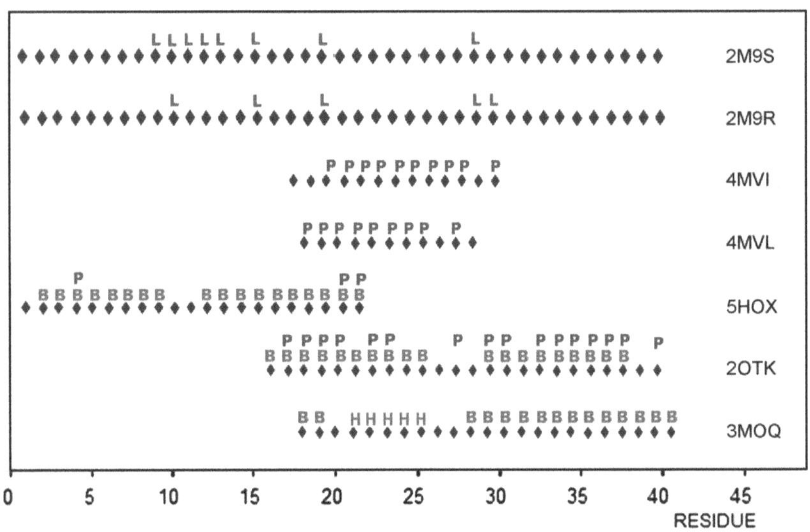

Fig. 9.B.13 Comparison of Aβ(1–42) fragments present in each analyzed structure. Specific ranges are indicated by the lower horizontal bars for each table entry. L – ligand binding, P – protein-peptide interactions, B – sheet, H – helix.

random coils. In the two former cases they also remain in full contact with external complexation partners. With regard to 2M9R and 2M9S, the presence of ligands appears to have a decisive influence on structural stability, likely due to steric hindrance (as a relatively large molecule the ligand effectively prevents complexation). In addition, by engaging residues which are quite far apart in the chain, the ligand forces it to adopt a random coil (and therefore non-amyloid-like) conformation.

The β-hairpin found in 5HOX (at 2–10) exhibits amyloid-like properties, while the fragment at 16–21 is characterized by high RD and high HvT, but also by high TvO. Consequently, it may be regarded as "weakly amyloid-like". Other fragments whose FOD status is similar to that found in amyloids include 22–28 in 2OTK and 17–22 in both 4MVL and 4MVI. As noted above, 2M9R and 2M9S do not appear to contain any amyloid-like fragments.

A common property of all presented forms of the Aβ(1–42) peptide is the need for an external factor which guides the folding process. The complexes described in this chapter were identified while seeking structural factors capable of arresting the growth of amyloids. In all cases, this is achieved by interaction with an external ligand or protein, producing a structure which does not permit unrestricted aggregation resulting in amyloid formation.

References

[1] Rauth S, Hinz D, Borger M, Uhrig M, Mayhaus M, Riemenschneider M, Skerra A. High-affinity Anticalins with aggregation-blocking activity directed against the Alzheimer β-amyloid peptide. Biochemical Journal 2016;473(11):1563–78. https://doi.org/10.1042/bcj20160114.

[2] Hoyer W, Grönwall C, Jonsson A, Ståhl S, Härd T. Stabilization of a β-hairpin in monomeric Alzheimer's amyloid-β peptide inhibits amyloid formation. Proceedings of the National Academy of Sciences 2008;105(13):5099–104. https://doi.org/10.1073/pnas.0711731105.

[3] Richard T, Papastamoulis Y, Waffo-Teguo P, Monti J-P. 3D NMR structure of a complex between the amyloid beta peptide (1–40) and the polyphenol ε-viniferin glucoside: implications in Alzheimer's disease. Biochimica et Biophysica Acta (BBA) - General Subjects 2013;1830(11):5068–74. https://doi.org/10.1016/j.bbagen.2013.06.031.

[4] Kreutzer AG, Hamza IL, Spencer RK, Nowick JS. X-ray crystallographic structures of a trimer, dodecamer, and annular pore formed by an aβ17–36 β-hairpin. Journal of the American Chemical Society 2016;138(13):4634–42. https://doi.org/10.1021/jacs.6b01332.

[5] Streltsov VA, Varghese JN, Masters CL, Nuttall SD. Crystal structure of the amyloid- p3 fragment provides a model for oligomer formation in alzheimer's disease. Journal of Neuroscience 2011;31(4):1419–26. https://doi.org/10.1523/jneurosci.4259-10.2011.

[6] Laskowski RA, Jabłońska J, Pravda L, Vařeková RS, Thornton JM. PDBsum: structural summaries of PDB entries. Protein Science 2017;27(1):129–34. https://doi.org/10.1002/pro.3289.
[7] Dułak D, Gadzała M, Banach M, Ptak M, Wiśniowski Z, Konieczny L, Roterman I. Filamentous aggregates of tau proteins fulfil standard amyloid criteria provided by the fuzzy oil drop (FOD) model. International Journal of Molecular Sciences 2018; 19(10):2910. https://doi.org/10.3390/ijms19102910.
[8] Dułak D, Banach M, Gadzała M, Konieczny L, Roterman I. Structural analysis of the Aβ(15–40) amyloid fibril based on hydrophobicity distribution. Acta Biochimica Polonica 2018. https://doi.org/10.18388/abp.2018_2647.

CHAPTER 10

Amyloids identification based on fuzzy oil drop model

Mateusz Banach, Irena Roterman
Department of Bioinformatics and Telemedicine, Jagiellonian University—Medical College, Krakow, Poland

Contents

Conceptual image showing the unrestricted capability for linear propagation of bands characterized by variable hydrophobicity (different degree of gray color). Dashed lines correspond to elongation beyond the boundaries of PDB structures.

The fuzzy oil drop model asserts the presence of a monocentric hydrophobic core described by a 3D Gaussian function. As previously shown, globular proteins tend to conform to this idealized distribution (with variable accuracy). In effect, the Gaussian describes an "ideal micelle", which, in accordance with information theory, contains very little information. This is due to its deterministic (symmetrical) structure, where the placement of all components can be predicted with high confidence (high probability means low information content).

Globular (or near-globular) proteins may be inscribed in 3D Gaussian capsules whose dimensions are adjusted to each case by manipulating their σ (sigma) coefficients. As previously discussed, such proteins may be treated as "intelligent micelles": in addition to adopting micelle-like conformations, they also encode specific information which is expressed as localized deviations from the theoretical distribution of hydrophobicity (note that such "improbable" structures may carry a substantial quantity of information).

From Globular Proteins to Amyloids
ISBN: 978-0-08-102981-7
https://doi.org/10.1016/B978-0-08-102981-7.00015-4

While there are many possible deviations from the monocentric model, one in particular is worth further analysis: we refer here to a strongly ordered system in which the hydrophobic core has been replaced by a linearly propagating sequence of "bands" (alternating between high and low hydrophobicity). This arrangement is observed in amyloids, and the present chapter provides arguments in support of defining an amyloid as a structure which exhibits this characteristic band-like pattern.

Another interesting property of amyloids is that rather than simply deviating from the Gaussian distribution, they may often be regarded as polar opposites thereof. This effect may be explained if we observe that in amyloid structures the actual distribution of hydrophobicity is determined by the intrinsic preferences of each participating residue, with no cooperative tendency to form a shared hydrophobic core (consistent with the Gaussian).

By assessing the specific discordance between theoretical (T) and observed (O) hydrophobicity distributions, we may formulate criteria for regarding a given structure as an amyloid.

Researchers often point to β-strands as being particularly prone to amyloid transformation. On the other hand, it can also be shown that amyloids may emerge from other types of structures — such as the tau protein [1]. Notably, the tau amyloid also satisfies the FOD-based amyloid identification criteria [2—4], which are presented below.

This chapter (10) is divided into three parts:

1. This part discusses the structure of amyloid forms of Aβ(1—42) chain fragments, based on PDB data.
2. In this part the Aβ(1—42) sequence is treated as a folding target (following the CASP procedure (http://predictioncenter.org)) in order to compare various alternative conformations for this polypeptide.
3. The specificity of sequence to generate characteristics secondary forms supporting globular of linear order is shown.

References

[1] Fitzpatrick AWP, Falcon B, He S, Murzin AG, Murshudov G, Garringer HJ, Scheres SHW. Cryo-EM structures of tau filaments from Alzheimer's disease. Nature 2017;547(7662):185—90. https://doi.org/10.1038/nature23002.

[2] Dułak D, Gadzała M, Banach M, Ptak M, Wiśniowski Z, Konieczny L, Roterman I. Filamentous aggregates of tau proteins fulfil standard amyloid criteria provided by the fuzzy oil drop (FOD) model. International Journal of Molecular Sciences 2018; 19(10):2910. https://doi.org/10.3390/ijms19102910.

[3] Dułak D, Banach M, Gadzała M, Konieczny L, Roterman I. Structural analysis of the Aβ(15–40) amyloid fibril based on hydrophobicity distribution. Acta Biochimica Polonica 2018. https://doi.org/10.18388/abp.2018_2647.
[4] Banach M, Konieczny L, Roterman I. Fuzzy oil drop model application—from globular proteins to amyloids. Computational Methods to Study the Structure and Dynamics of Biomolecules and Biomolecular Processes 2018:639–58. https://doi.org/10.1007/978-3-319-95843-9_19.

CHAPTER 10.A

Amyloid as a ribbon-like micelle

Mateusz Banach[1], Irena Roterman[1]
[1]Department of Bioinformatics and Telemedicine, Jagiellonian University—Medical college, Krakow, Poland

Contents

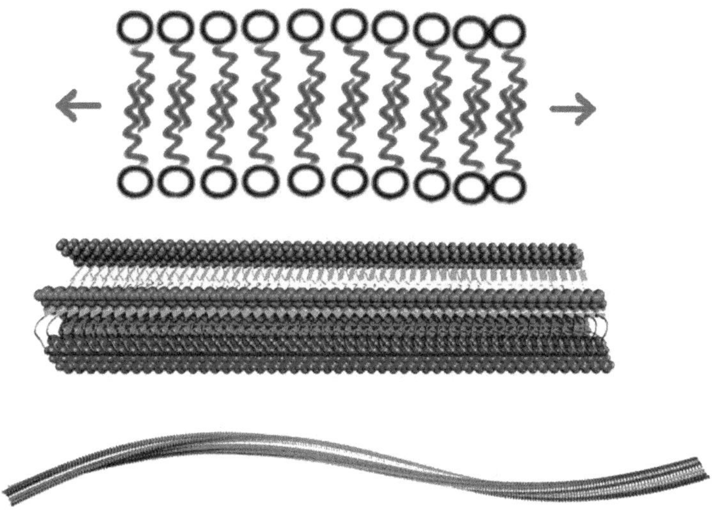

Different scales of ribbon-like micelle presentation visualizing the linear propagation of hydrophilic (white circles) and hydrophobic (black zigzag lines) bands (top). The central picture visualizes the amyloid Aβ(15—40) containing 50 polypeptides. The long fragment of fibril Aβ(15—40) containing 200 peptides (bottom). The

From Globular Proteins to Amyloids
ISBN: 978-0-08-102981-7
https://doi.org/10.1016/B978-0-08-102981-7.00016-6

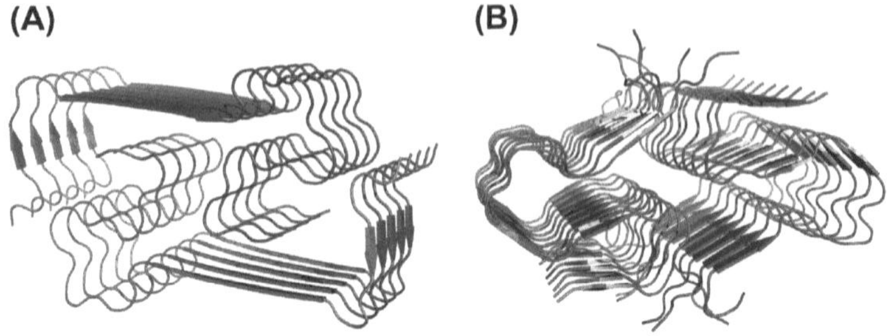

Fig. 10.A.1 3D presentation superfibrils. (A) Aβ(1−40) (2MVX). (B) Aβ(11−42) (5KK3). An example chain in each protein is distinguished from the rest of the complex by red and green colors. Red color marks 11-40 fragments where sequences of the proteins are identical. Conversely, green color denotes fragments where the sequences differ.

structures generated as the multiplication of short fragments available in PDB. The intensity of gray color is proportional to hydrophobicity level.

In this chapter we will discuss the following amyloid proteins:

1. Amyloid Aβ(1−40) peptide with the Osaka mutation (E22Δ) (PDB ID: 2MVX) [1].
2. Amyloid Aβ(11−42) fragment 672−713 −superfibril consisting of Aβ42 molecules, each containing four β-strands in an S-shaped configuration (PDB ID: 5KK3) [2].
3. Prion domain of the fungal prion HET-s in its amyloid form, available in PDB as 2KJ3 [3].

Two of these structures (2MVX and 5KK3) are fragments of the Aβ(1−42) amyloid. Despite substantial sequential similarities (identical sequence at 11−40), they exhibit major conformational differences (Fig. 10.A.1). Both proteins share a common property: they form superfibrils which consist of two dimerized protofibrils each, although with differing symmetry: the structure of 2MVX includes five-chain protofibrils, while in the case of 5KK3 each protofibril is composed of 10 chains.

Amyloid Aβ(1−40) peptide with the Osaka mutation (E22Δ)

The theoretical distribution for Aβ(1−40) (2MVX) (Fig. 10.A.2) accounts for an expected peak at the protein's core. Here, T consists of multiple profiles, characterized by increasing values of hydrophobicity − an

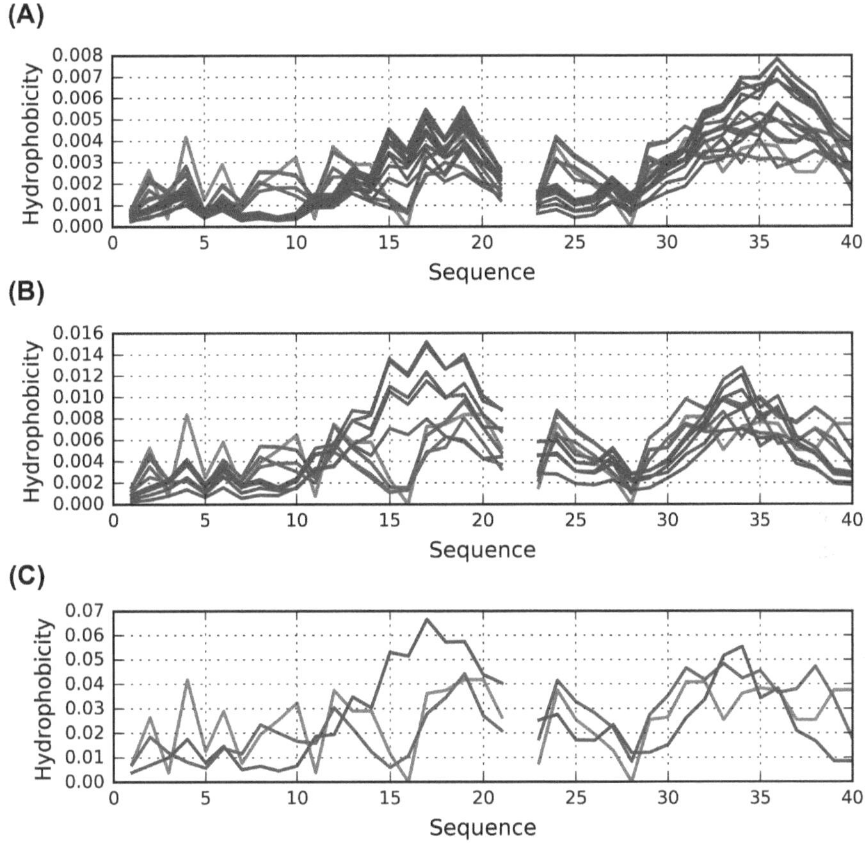

Fig. 10.A.2 Theoretical (T, blue), observed (O, red) and intrinsic (H, green) hydrophobicity distribution profiles for Aβ(1−40) (2MVX). (A) superfibril. (B) protofibril (chains A, B, C, D, E). (C) individual (central) chain of protofibril (C). In each case a Gaussian capsule is constructed specifically for the structural unit undergoing analysis. There is no residue number 22 in the sequence, however there is no gap in the chains within the PDB structure.

effect caused by the variable distance of successive polypeptides from the center of the fibril (where high hydrophobicity is expected).

In contrast, observed (O) distribution plots reveal near-identical values for all internal chains, with only a slight decrease in hydrophobicity in the outlying (edge) chains. If we note that experimentally observed fibrils are capable of unlimited propagation, it follows that these internal chains are representative of the distribution of hydrophobicity which spans the entire fibril. Differences between the theoretical and observed distribution −

particularly concerning the location of hydrophobicity minima and maxima — are stark. Unlike T, where only two peaks are present, the observed distribution resembles a high-frequency sinusoid and comprises five distinct maxima.

Calculation of fuzzy oil drop parameters (Table 10.A.1) shows that both the superfibril and the protofibril exhibit high RD values in both variants

Table 10.A.1 Fuzzy oil drop parameters for Aβ(1–40) (2MVX), computed for superfibril, protofibril and chain C, both as a standalone structure, as part of the protofibril and as part of the superfibril. Values given in bold represent the status interpreted as amyloid seed.

Aβ(1–40)		RD		Correlation coefficient		
(2MVX)	Fragment	T-O-R	T-O-H	HvT	TvO	HvO
Superfibril		0.590	0.592	0.438	0.674	0.727
Chain C in superfibril						
Complete	1–40	0.608	0.620	0.459	0.665	0.784
	1–10	0.663	0.569	0.401	0.216	0.733
	11–16	**0.706**	**0.475**	**−0.420**	**−0.496**	**0.910**
	17–28	0.564	0.527	0.707	0.554	0.927
	29–40	0.853	0.648	0.250	0.298	−0.003
	10–27	0.698	0.646	0.250	0.221	0.886
Chain C in protofibril						
Complete	1–40	0.632	0.666	0.312	0.369	0.784
	1–10	0.554	0.532	0.692	0.333	0.693
	11–16	**0.690**	**0.475**	**−0.413**	**−0.669**	**0.892**
	17–28	0.545	0.526	0.735	0.552	0.923
	29–40	0.816	0.704	0.177	0.688	0.194
	10–27	0.691	0.660	0.162	0.020	0.872
Chain C as individual unit						
Complete	1–40	0.636	0.562	0.295	0.363	0.616
	1–10	0.738	0.541	0.555	−0.287	0.257
	11–16	**0.743**	**0.336**	**−0.464**	**−0.720**	**0.693**
	17–28	0.621	0.413	0.749	0.455	0.822
	29–40	0.797	0.690	0.159	0.611	−0.022
	10–27	0.695	0.507	0.195	0.101	0.697

(T-O-R and T-O-H). Notably, the alignment between O and H is better than between O and R, indicating that formation of the fibril is driven by the intrinsic hydrophobicity of its constituent residues. This view is further supported by high values of HvO, especially when compared to the other two correlation coefficients (even though the dominance of HvO is not overwhelming).

The status of the C chain, which occupies a central position and therefore provides the best representative for a putative "endless" fibril, indicates further deviation from T in favor of H. This conclusion follows from analysis of its RD and correlation coefficients. The RD for T-O-H is below 0.5, however the very large bias in correlation coefficient significantly favoring the relation HvO suggests this interpretation.

Individual chains may be analyzed in the context of their respective super- or protofibrils, but also on their own (i.e. using a 3D Gaussian capsule which is limited to the chain undergoing analysis). In this case, the status of the 11–16 fragment is particularly noteworthy since it exhibits negative HvT and TvO values while its HvO value is particularly high. Negative correlation indicates that the observed distribution is not only discordant from theoretical values, but in fact opposes it, at least to a certain extent. Clearly, intrinsic hydrophobicity has a particularly strong effect on O within this fragment. The remaining fragments (corresponding to each local maximum) also reveal close alignment between the intrinsic and observed distributions of hydrophobicity.

In effect, the following parameters may be regarded as characteristic "markers" of an amyloid structure: high values of RD (both variants), negative values of HvT and TvO, and a strongly positive value of HvO (Fig. 10.A.2 and Table 10.A.1).

The parameters listed in Table 10.A.1 show that the 11–16 fragment is particularly discordant from the monocentric distribution of hydrophobicity. Lys16 is an especially noteworthy residue, since it acts as a "breaker", splitting the expected broad maximum into two shorter (and less hydrophobic) fragments. Given that the same arrangement is replicated in successive chains, we may assume that residue 16 acts as a hydrophilic band interposed between two adjacent hydrophobic bands. This linear pattern emerges as a result of clustering between identical chains along the fibril's axis of propagation. As a result, the fibril as a whole is dominated by alternating bands of high and low hydrophobicity which depend upon the intrinsic properties of individual residues (Fig. 10.A.3).

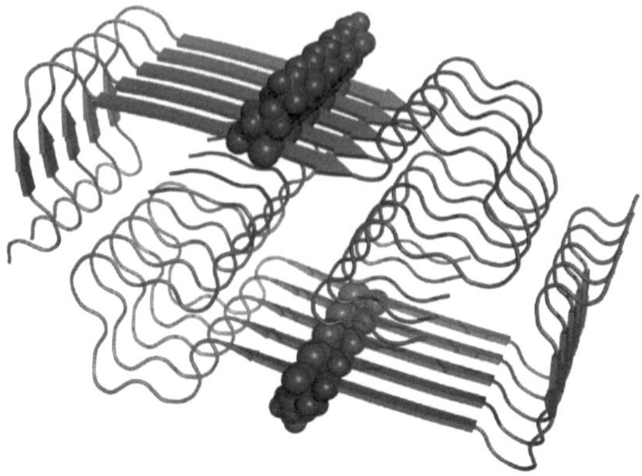

Fig. 10.A.3 3D presentation of Aβ(1−40) (2MVX), with 11−16 fragment in each chain highlighted in red. Lys16, located at the end of this fragment, is shown as blue spheres.

Aβ(11−42) amyloid

When a similar analysis is performed for the Aβ(11−42) amyloid (5KK3), it turns out that the aforementioned criteria are met by the superfibril, by each protofibril as well as by individual chains. In this case, however, the fragment which satisfies the stated conditions is the one at 24−28 (Fig. 10.A.4, Table 10.A.2). A local maximum is present where the theoretical model expects a local minimum (if the molecule were to adopt a globular conformation). This also implies that a locally hydrophobic band is exposed to the outside environment − an unfavorable condition, particularly if the band stretches along the entire axis of the fibril. It is important to recall that each local maximum observed in an individual chain translates into a long band, and that the overall distribution of hydrophobicity in the amyloid fibril is very much unlike the distribution predicted for a globular protein (Fig. 10.A.5).

Prion amyloid

The next structure we discuss is prion amyloid (2KJ3). Three chains, 79 residues each, form a solenoid which consists of six fragments (two per chain). The shape of the solenoid resembles a double "C", with an outer part and an inner part (Fig. 10.A.6). This amyloid differs from previous examples due to its origin and sequence. What is more, fragments which

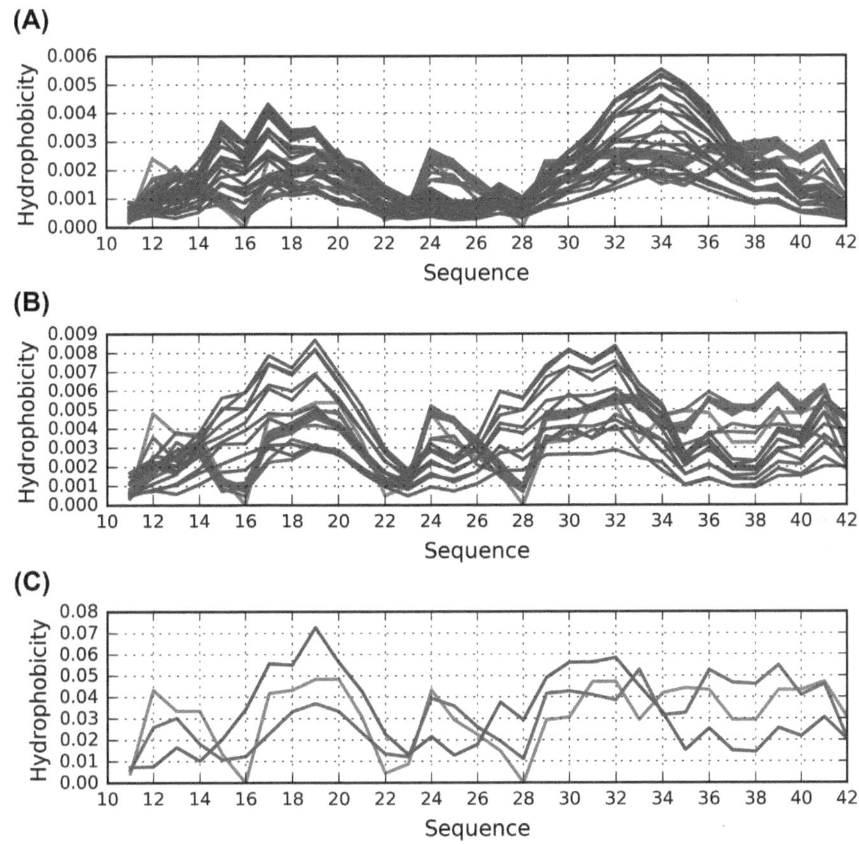

Fig. 10.A.4 Theoretical (T, blue), observed (O, red) and intrinsic (H, green) hydrophobicity distribution profiles for Aβ(11−42) (5KK3). (A) superfibril. (B) protofibril (chains A, B, C, D, E, F, G, H, I). (C) individual (central) chain of protofibril (E). In each case a Gaussian capsule is constructed specifically for the structural unit undergoing analysis.

form the amyloid fibril are not sequentially identical (the sequence of outer "C" differs from the sequence of inner "C"). Note that while amyloids constructed from identical fragments enable rapid clustering of local maxima and minima, the same is not necessarily true for structures which comprise different local sequences.

Linear propagation of alternating hydrophobicity bands appears to be an obvious consequence of alignment of identical fragments, however, in the presented case, the underlying mechanism is somewhat different. Within the solenoid each pair of adjacent twists is formed by two distinct fragments which do not share the same sequence. Aβ(11−42) (2KJ3) therefore appears to be an interesting case study, showing how the formation of alternating bands may occur even in the absence of sequential identity.

Table 10.A.2 Fuzzy oil drop parameters for 5KK3, computed for superfibril, protofibril and chain E, both as a standalone structure, as part of the protofibril and as part of the superfibril. Values given in bold represent the status interpreted as amyloid seed. Values underlined distinguish the status close to amyloid-like expressing significant bias with very high correlation coefficient for HvO relation.

Aβ(11−42)		RD		Correlation coefficient		
(5KK3)	**Fragment**	**T-O-R**	**T-O-H**	**HvT**	**TvO**	**HvO**
Superfibril		0.620	0.534	0.330	0.440	0.756
Protofibril		0.608	0.622	0.235	0.335	0.750
Chain E						
Superfibril		0.565	0.594	0.395	0.466	0.782
Protofibril I		0.569	0.600	0.299	0.286	0.784
Individual E		0.660	0.555	0.355	0.263	0.698
Chain E in superfibril						
Complete	11−42	0.565	0.595	0.396	0.467	0.783
	11−16	0.514	0.296	−0.382	0.265	0.651
	17−23	0.225	0.416	0.826	0.811	0.988
	24−28	**0.575**	0.454	**−0.050**	**−0.215**	**0.982**
	29−35	0.964	0.708	0.423	0.487	0.119
	36−42	0.774	0.784	0.060	0.522	0.530
Chain E in protofibril						
Complete	11−42	0.570	0.601	0.300	0.286	0.784
	11−16	0.538	0.302	−0.376	0.057	0.762
	17−23	0.088	0.189	0.952	0.952	0.986
	24−28	**0.715**	**0.607**	**−0.774**	**−0.872**	**0.982**
	29−35	0.789	0.447	−0.193	0.799	−0.217
	36−42	0.741	0.813	0.647	0.064	0.687
Chain E as individual unit						
Complete	11−42	0.660	0.555	0.356	0.263	0.699
	11−16	0.714	0.345	−0.572	−0.191	0.850
	17−23	0.187	0.193	0.951	0.932	0.931
	24−28	**0.752**	0.410	**−0.556**	**−0.683**	**0.976**
	29−35	0.798	0.478	−0.156	0.526	−0.637

On the other hand, the presented amyloid also presents a significant drawback — from the point of view of our study — namely, it is very short, consisting of only three chains (i.e. six solenoid folds). Under these conditions it is difficult to obtain proof of unrestricted propagation given that

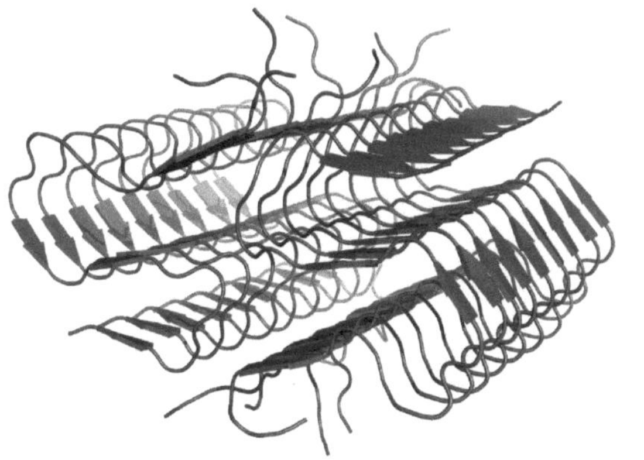

Fig. 10.A.5 3D presentation of Aβ(11–42) (5KK3), with 24–28 fragment highlighted in red.

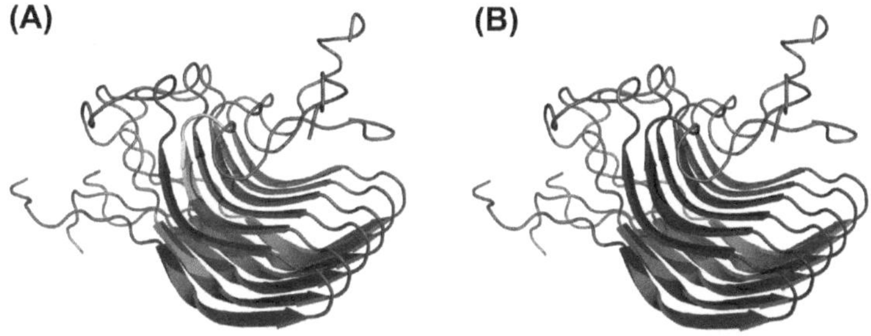

Fig. 10.A.6 3D presentation of prion amyloid (2KJ3). (A) example chain distinguished from the rest of the complex by colors: blue – fragment 241–249, green – fragment 257–264, yellow – fragment 279–283 (rest of the chain is shown in red). (B) solenoid: outer "C" shown in red (fragments 224–235, 261–271 of every chain) and inner "C" shown in blue (fragments 235–246, 271–283 of every chain).

the only non-outlying position is occupied by chain (B) Consequently, RD parameters calculated in this instance (Table 10.A.3) are somewhat less unequivocal than in our previous examples. While intrinsic hydrophobicity is clearly the driving force, shaping the tertiary conformation of 2KJ3 (note the high value of HvO), the solenoid itself is accordant with the theoretical distribution of hydrophobicity – which would suggest the presence of a monocentric hydrophobic core (Fig. 10.A.7).

Table 10.A.3 Fuzzy oil drop parameters for prion amyloid (2KJ3), computed for fibril, chains (treated as part of the fibril) and their fragments. Values given in bold represent the status interpreted as amyloid seed. Values underlined – status of near amyloid form due to the biased relation of correlation coefficients.

Prion amyloid (2KJ3)	Fragment	RD T-O-R	RD T-O-H	Correlation coefficient HvT	Correlation coefficient TvO	Correlation coefficient HvO
Fibril		0.535	0.411	0.143	0.555	0.615
Chains in fibrl						
Chain A		0.564	0.437	0.101	0.427	0.585
Chain B		0.485	0.368	0.149	0.606	0.615
Chain C		0.534	0.402	0.200	0.592	0.665
Solenoid						
Chain A + B + C	224–246 + 261–283	0.442	0.373	0.373	0.706	0.743
Chain A	241–249	0.846	0.725	−0.047	−0.261	0.340
	257–264	**0.649**	**0.700**	**−0.378**	**−0.410**	**0.781**
	279–283	0.513	0.047	0.645	−0.947	−0.658
Chain B	241–249	0.787	0.730	0.066	0.451	0.662
	257–264	0.652	0.464	-0.106	−0.466	0.612
	279–283	**0.803**	0.346	**-0.688**	**−0.940**	**0.893**
Chain C	241–249	0.836	0.808	0.149	0.417	0.723
	257–264	0.461	0.453	0.174	0.350	0.581
	279–283	**0.864**	0.291	**−0.684**	**−0.952**	**0.578**

The hydrophobicity distribution plots (Fig. 10.A.7) reveal differentiation of individual chains, which is caused by the small size of the amyloid (limited to only three chains). Highlighted fragments deviate from the theoretical distribution in a similar way to our previous examples (i.e. they remain in "active" opposition to T). Much like other proteins discussed in this chapter, 2KJ3 includes "breakers" between strongly hydrophobic bands, and presents local maxima where T expects hydrophobicity to remain low. All these effects counteract the formation of a shared hydrophobic core.

When considering the complex as a whole, the computed value of RD indicates that O remains consistent with T. This, however, may be attributed to the small size of the molecule, as well as to loose fragments which mediate contact with the environment and are capable of adopting a micellar pattern. As shown in Table 10.A.3, individual fragments (in each of the three chains) clearly exhibit amyloid-like properties. This is particularly true for

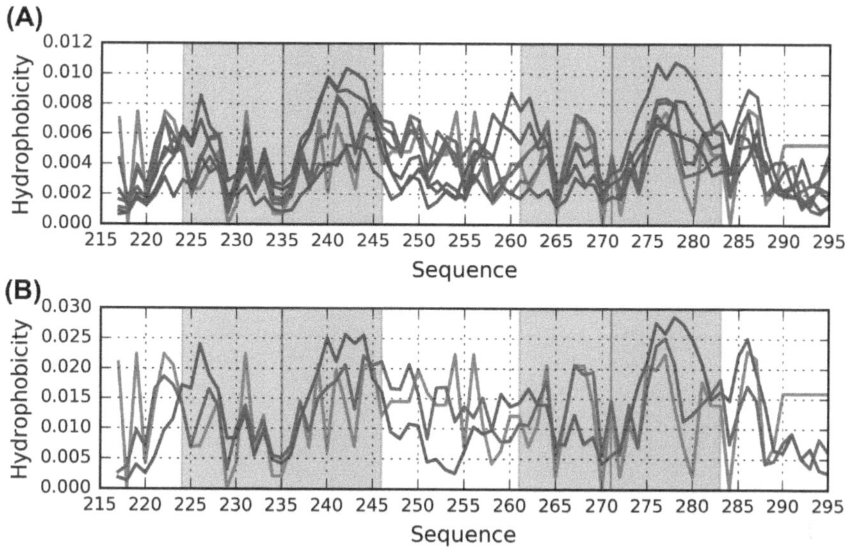

Fig. 10.A.7 Theoretical (T, blue), observed (O, red) and intrinsic (H, green) hydrophobicity distribution profiles for prion amyloid (2KJ3). (A) fibril. (B) chain B (central). In each case a Gaussian capsule is constructed specifically for the structural unit undergoing analysis. Background colors mark the locations of solenoid components: outer "C" (red - fragments 224-235, 261-271) and inner "C" (blue - fragments 235-246, 271-283).

the fragment at 279–283. There is also a notable tendency for bands of high and low hydrophobicity to propagate in an alternating fashion (Fig. 10.A.8).

Fig. 10.A.8 reveals bands of hydrophobicity which emerge despite sequential differences. Notably, local maxima are separated by a deep minimum in each fragment and individual fragments remain in closeness, proximity to one another even though their sequences are not identical. The residue at position 6 in each profile (position in the common regularly ordered fragments) provides a characteristic "breaker" (position 243 and 280), splitting the expected local maximum into two smaller maxima, in a manner similar to Aβ(1–40) (2MVX). This band of low hydrophobicity enables the structure to propagate in a linear manner, thereby "actively" opposing the expected distribution. The effect is not local in scope, but instead dominates the entire fibril. The role of hydrophobicity in amyloid formation is underscored by the observation that only hydrophobic forces may cause tight clustering of strongly charged residues – in contrast to electrostatic interactions which disfavor such structural alignment.

When analyzing the observed distributions we may note that local maxima appearing in successive fragments of the chain differ in amplitude (Fig. 10.A.9). This effect is caused by differences in the sequences of individual fragments; however it does not override the overall conformational

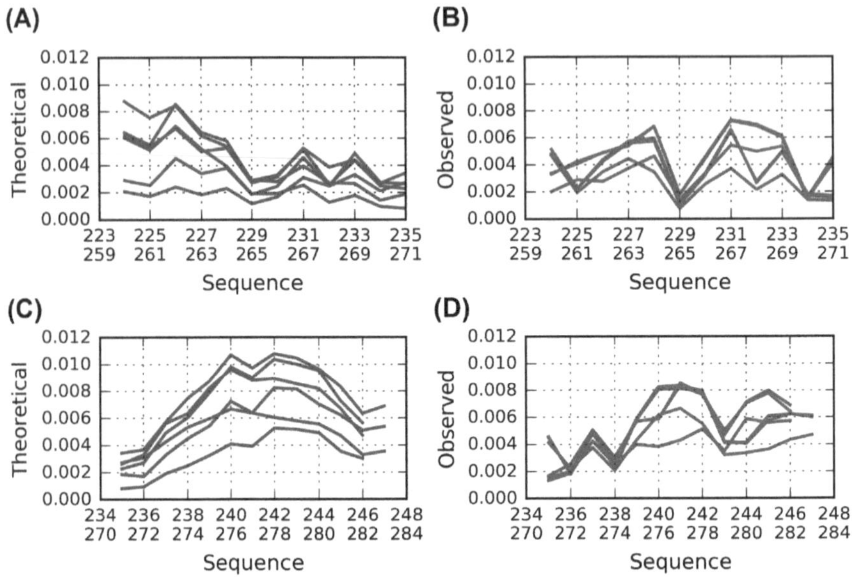

Fig. 10.A.8 Theoretical (T, blue) and observed (O, red) hydrophobicity distribution profiles for prion protein (2KJ3) fragments comprising the solenoid. (A) theoretical distribution in outer "C" (fragments 224–235, 261–271). (B) observed distribution in outer "C" (fragments 224–235, 261–271). (C) theoretical distribution in inner "C" (fragments 235–246, 271–283). (D) observed distribution in inner "C" (fragments 235–246, 271–283).

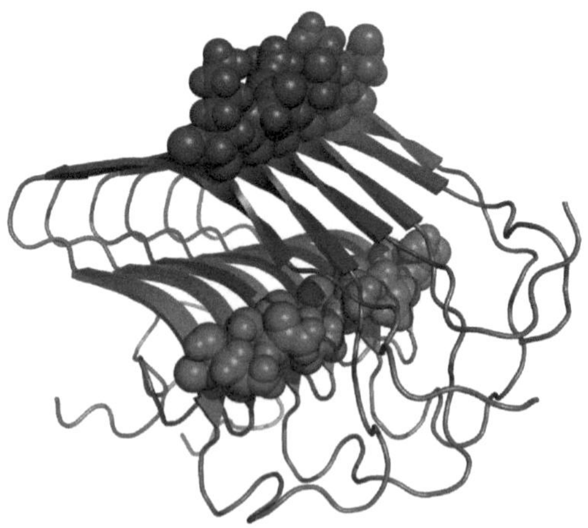

Fig. 10.A.9 3D presentation of prion amyloid (2KJ3) distinguishing fragments 224–235, 261–271 (blue) and 235–246, 271–283 (red). Spheres highlight the positions of residues 229, 265 (blue) and 243, 279 (red).

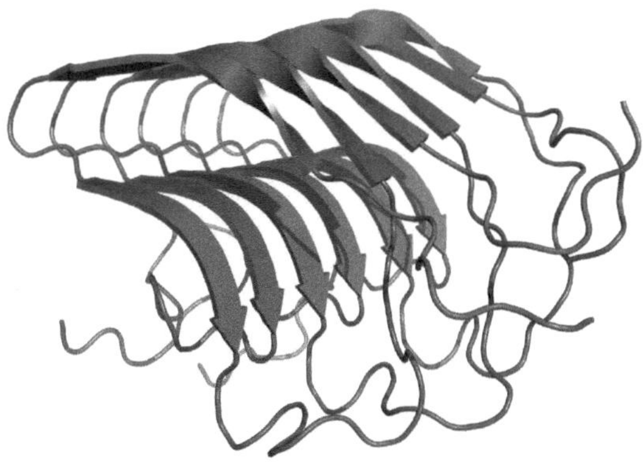

Fig. 10.A.10 3D presentation of prion amyloid (2KJ3) showing fragments which satisfy the proposed amyloid identification criteria. The fragments distinguished as red were identified using the criteria: negative correlation coefficients for HvT and TvO relations accompanied by high values for HvO relation. These fragments are as follows: chain A – 219–226, 239–245, 273–282; chain B – 219–226, 237–245, 273–282; chain C – 219–226, 237–245, 276–282.

pattern, where two local maxima and – even more importantly – the interposing local minimum propagate in a linear fashion.

Comparing T and O distributions for the listed fragments reveals significant differences. While replication of local minima and maxima is not as evident as in the case of Aβ(1–42), the tendency is nevertheless clear and pervades the structure of the amyloid.

As already noted, the limited length of 2KJ3 conceals its fibrillary properties (unlike in previously discussed amyloids which involve greater numbers of individual chains).

Fig. 10.A.10 shows that some of the fragments which exhibit negative correlation coefficients (5 aa moving frame analysis) including also these which do not contribute to the solenoid. In summary, it should be noted that detection and characterization of amyloid structures may be based on parameters provided by the fuzzy oil drop model – in particular, high values of RD (in both variants – T-O-R and T-O-H), negative values of HvT and TvO, along with high values of HvO. These conditions are exemplified by various amyloids listed in PDB, including Aβ(1–40) (2MXU), Aβ(15–40) (2MPZ) [4] and the tau amyloid [5]. It therefore seems justifiable to extend our conclusions to other amyloid structures.

If we accept the premise that the proposed criteria unambiguously categorize certain protein structures as amyloid-like, the following conclusion can be formulated: The amyloid is nothing more than a micelle — specifically, a ribbonlike micelle. Much like a spherical micelle, the ribbonlike micelle is a self-determining form. If the conformation of the initial chain in the sequence is known with precision, the structure of all subsequent chains may easily be computed. This deterministic property means that, as a whole, the ribbonlike amyloid carries no additional information, regardless of its length.

In silico experiment

In order to assess the ability of the Aβ(1−40) (2MVX) to adopt other conformations, an *in silico* experiment was performed. The polypeptide was subjected to folding simulations using Robetta [6,7] and I-Tasser [8,9], recognized as the most accurate in recent editions of the CASP challenge [10]. The experiment also involved a FOD-based simulation which promotes the formation of a monocentric hydrophobic core and therefore presents an interesting alternative to approaches which do not acknowledge hydrophobic effects [11,12].

Results of the experiment are presented and discussed in Chapter 10.B.

References

[1] Schütz AK, Vagt T, Huber M, Ovchinnikova OY, Cadalbert R, Wall J, Meier BH. Atomic-resolution three-dimensional structure of amyloid β fibrils bearing the Osaka mutation. Angewandte Chemie International Edition 2014;54(1):331−5. https://doi.org/10.1002/anie.201408598.

[2] Colvin MT, Silvers R, Ni QZ, Can TV, Sergeyev I, Rosay M, Griffin RG. Atomic resolution structure of monomorphic Aβ42 amyloid fibrils. Journal of the American Chemical Society 2016;138(30):9663−74. https://doi.org/10.1021/jacs.6b05129.

[3] Van Melckebeke H, Wasmer C, Lange A, Eliso AB, Loquet A, Bockmann A, Meier BH. Atomic-resolution three-dimensional structure of HET-s(218−289) amyloid fibrils by solid-state NMR spectroscopy. Journal of the American Chemical Society 2010;132(39):13765−75. https://doi.org/10.1021/ja104213j.

[4] Dułak D, Banach M, Gadzała M, Konieczny L, Roterman I. Structural analysis of the Aβ(15−40) amyloid fibril based on hydrophobicity distribution. Acta Biochimica Polonica 2018. https://doi.org/10.18388/abp.2018_2647.

[5] Dułak D, Gadzała M, Banach M, Ptak M, Wiśniowski Z, Konieczny L, Roterman I. Filamentous aggregates of tau proteins fulfil standard amyloid criteria provided by the fuzzy oil drop (FOD) model. International Journal of Molecular Sciences 2018;19(10): 2910. https://doi.org/10.3390/ijms19102910.

[6] Kim DE, Chivian D, Baker D. Protein structure prediction and analysis using the Robetta server. Nucleic Acids Research 2004;32(Web Server):W526−31. https://doi.org/10.1093/nar/gkh468.

[7] http://robetta.bakerlab.org.

[8] Zhang Y. I-TASSER server for protein 3D structure prediction. BMC Bioinformatics 2008;9(1):40. https://doi.org/10.1186/1471-2105-9-40.

[9] https://zhanglab.ccmb.med.umich.edu/I-TASSER.

[10] http://predictioncenter.org.
[11] Fitzpatrick AWP, Falcon B, He S, Murzin AG, Murshudov G, Garringer HJ, Scheres SHW. Cryo-EM structures of tau filaments from Alzheimer's disease. Nature 2017;547(7662):185–90. https://doi.org/10.1038/nature23002.
[12] Banach M, Konieczny L, Roterman I. Fuzzy oil drop model application—from globular proteins to amyloids. Computational Methods to Study the Structure and Dynamics of Biomolecules and Biomolecular Processes 2018:639–58. https://doi.org/10.1007/978-3-319-95843-9_19.

CHAPTER 10.B

Analysis of alternative conformations of the Aβ(1−40) amyloid protein

Dawid Dułak[1], Małgorzata Gadzała[2], Mateusz Banach[3], Irena Roterman[3]

[1]Department of Biophysics, Faculty of Physics, Astronomy and Applied Computer Sciencee Jagiellonian University, Krakow, Poland
[2]Cyfronet AGH Academic Computer Centre CYFRONET - University of Science and Technology in Cracow, Kraków, Poland; Currently - Schibsted Tech Polska Sp. z o. o., Kraków, Poland
[3]Department of Bioinformatics and Telemedicine, Jagiellonian University—Medical College, Krakow, Poland

Contents

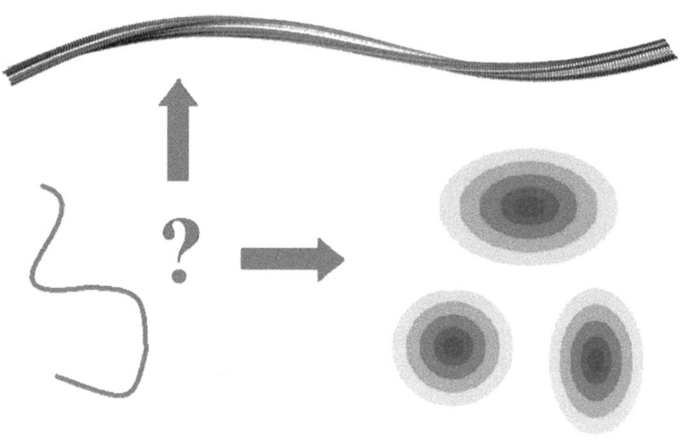

From Globular Proteins to Amyloids
ISBN: 978-0-08-102981-7
https://doi.org/10.1016/B978-0-08-102981-7.00017-8

Making decision by the folding polypeptide: amyloid ? or globular ?

Alternative structural forms of polypeptide Aβ(1−40) polypeptide which includes the Osaka mutation

The amyloid structures like for example Aβ(1−40) [1] become available due to new technique ssNMR [2]. A polypeptide corresponding to the Aβ(1−40) sequence with the so-called Osaka mutation was subjected to an *in silico* experiment. Specialized folding software was used to generate five different folds per software package, in accordance with the rules of the CASP challenge. The experiment provided structures which could be regarded as alternative with respect to the experimentally observed conformation (PDB ID: 2MVX [1]). This, in turn, facilitated comparative analysis aimed at identifying fragments which support or deviate from the expected monocentric distribution of hydrophobicity. The presence of a centralized hydrophobic core provides the protein with solubility and thereby prevents unchecked complexation (potentially producing an amyloid). In addition to the above, the folding process was simulated in the presence of an external force field (FOD model) which mimics the active participation of aqueous solvent in folding process. The resulting structures provide evidence that — under the appropriate conditions — the Aβ(1−40) polypeptide may adopt a globular conformation, and suggest that the environment plays a critical role in this process.

In an attempt to identify the causes of amyloid transformation of sequences which include the Aβ(1−40) fragment, we have performed an *in silico* experiment to identify possible conformational preferences of this polypeptide. The experiment involved folding simulation tools currently regarded as the most accurate: I-Tasser [3−6] and Robetta [7−9] - two of the highest-scoring CASP challenge entrants [10]. Both are capable of predicting the conformation of chains with a given sequence, and each produces five distinct candidate structures, referred to as models. We subjected these models to comparative analysis set against the backdrop of the experimentally observed structure (obtained using ssNMR). In addition to the above, the input polypeptide was also subjected to simulations based on the fuzzy oil drop (FOD) model, which asserts the presence of an external force field representing the aqueous solvent and treated as a continuum — unlike other algorithms where the solvent is modeled as a collection of individual molecules [11,12]. In FOD, the solvent is mathematically represented by a 3D Gaussian form, directing hydrophobic residues toward the center of the protein body while exposing hydrophilic residues on its surface. The resulting set of models enhances our ability to perform comparative analysis of amyloid structures with the outcomes of *in silico* folding simulations.

All computations using FOD were performed at the Academic Computing Center CYFRONET AGH using resources provided by the PL-Grid infrastructure.

The resulting set of 20 models (FOD: 5, I-Tasser: 5, Robetta: 10) is the subject of the presented analysis. Each model will be compared to the experimentally determined target structure − both in the context of the superfibril and as an individual chain.

In order to carry out comparative analysis, we begin by computing fuzzy oil drop coefficients for all models. This includes values of RD (relative distance − as defined by Kullback-Leibler divergence entropy) in two distinct reference models: T-O-R and T-O-H respectively [13].

The FOD model, presented in Chapter 1, suggests that the value of RD(T-O-H) is particularly important. As proposed in Refs. [14,15], which deals with the structure of the tau amyloid, amyloid seeds may be identified by looking for high values of RD (T-O-R as well as T-O-H), along with negative values of HvT and TvO and strongly positive values of HvO. These specific conditions indicate that the given fragment opposes the theoretical distribution of hydrophobicity, and is dominated by the intrinsic properties of its residues.

Values of RD and correlation coefficients may be calculated for the entire protein (polypeptide), but also for selected fragments. In the latter case, the process tells us whether the given fragment contributes to the creation of a hydrophobic core or opposes it

The structure listed in PDB under ID 2MVX consists of two protofibrils exhibiting C2 symmetry. Each protofibril resembles a flattened "C", and contacts the opposing protofibril at both tips, while exposing its backside to the environment. In order to determine the FOD status of the superfibril we compute its T and O distributions (Figs. 10.B.1 and 10.B.2). As shown in Fig. 10.B.1, these distributions are a poor match for each other. In particular, no concentration of hydrophobicity can be

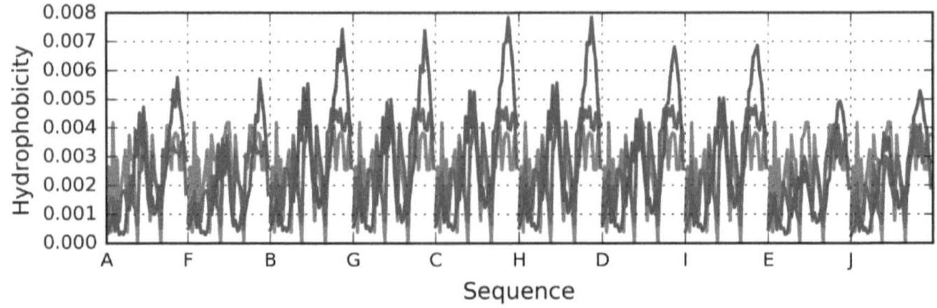

Fig. 10.B.1 Theoretical (T, blue), observed (O, red) and intrinsic (H, green) hydrophobicity distribution profiles for Aβ(1−40) (2MVX) superfibril.

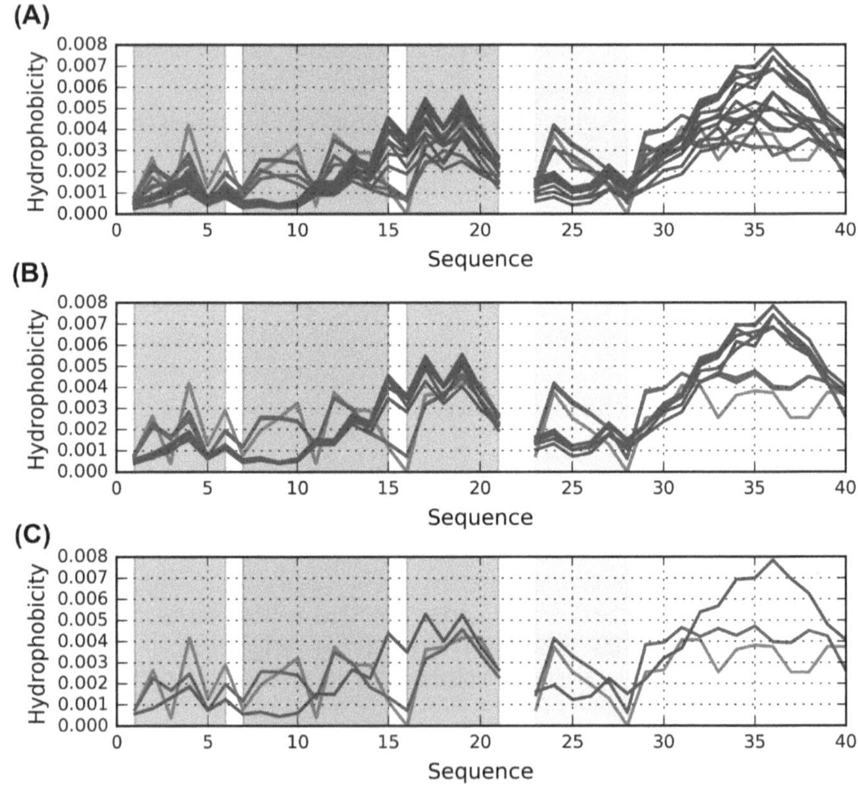

Fig. 10.B.2 Theoretical (T, blue), observed (O, red) and intrinsic (H, green) hydrophobicity distribution profiles for Aβ(1−40) (2MVX). (A) superfibril (chains A − J). (B) central part of the superfibril (following elimination of the outlying chains A, E, F and J). (C) individual (central) chain C treated as part of superfibril. Colored backgrounds highlight fragments analyzed in this study: red − 1-6, blue − 7-15, green − 16-21, yellow − 23-28. There is no residue number 22 in the sequence, however there is no gap in the chains within the PDB structure.

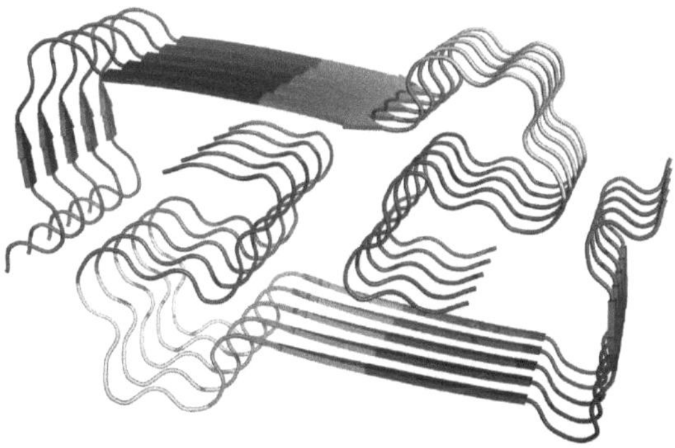

Fig. 10.B.3 3D presentation of Aβ(1–40) (2MVX) with fragments analyzed in this study highlighted by different colors: red – 1–6, blue – 7–15, green – 16–21, yellow – 23–28. Rest of each chain (29–40) is gray.

observed in the central part of the structure; instead, hydrophobicity is evenly distributed along the fibril's main axis. This type of distribution, where bands of high and low hydrophobicity propagate in an alternating fashion, resulting in a sinusoid pattern, is typical for amyloids. It follows from the repetitive nature of the input chain, as well as from conformational symmetries between each set of fragments making up the fibril (Fig. 10.B.3).

In the presented case, FOD parameters adopt the following values: RD (T-O-R) = 0.590; RD (T-O-H) = 0.592; HvT = 0.438; TvO = 0.673; HvO = 0.727. These results suggest that the structure as a whole does not contain a prominent hydrophobic core, and furthermore that the superfibril represents a consensus between the tendency to generate such a core and the intrinsic properties of each participating residue.

In contrast to Fig. 10.B.2A, profiles shown in Fig. 10.B.2B represents the distribution of hydrophobicity taking the entire superfibril as the structural unit for 3D Gauss construction.

Theoretical distribution, plotted in Fig. 10.B.2A and Fig. 10.B.2B, reveal the expected concentration of hydrophobicity in the central section of the fibril. The variability observed in Fig. 10.B.2B is due to the presence of edge chains, which lack an outlying neighbor and therefore exhibit slightly lower hydrophobicity than their centrally located counterparts (Fig. 10.B.2B). In contrast, the intrinsic distribution (H) distribution follows a sinusoidal pattern since all participating chains share identical sequences. The observed distribution (O) is also sinusoidal, consisting of alternating minima and maxima. Fig. 10.B.2B also reveals the high similarity of repeating patterns for O and H.

Fig. 10.B.2B visualizes the theoretical and observed distributions for the central chain treated as part of superfibril. Comparing both plots reveals significant differences at positions 7–15, 16–21, 22–27 and 23–28. Rather than being limited to modest deviations, these differences point to an entirely different structural pattern, counteracting the tendency to form a shared hydrophobic core.

Similar discordance is observed when calculating hydrophobicity profiles for chain C treated as part of protofibril (centrally placed chain C – Fig. 10.B.2C). The abovementioned discordant fragments will be subjected to further analysis, comparing them with the structural properties of *in silico* models.

Plotting a 3D Gaussian for the entire complex (superfibril) also enables us to determine the status of interface fragments, i.e. those residues which remain in contact with adjacent protofibrils. The status of the interface (residues satisfying the distance criterion according to PDBsum [16]: 3, 4,15, 28, 29, 30, 37–40) expressed by FOD parameters is given by the following RD values: 0.432 and 0.387 (T-O-R and T-O-H respectively), while correlation coefficients are 0.378, 0.672 and 0.658 (HvT, TvO and HvO respectively). These values indicate good alignment between the observed distribution of hydrophobicity and the distribution predicted by the fuzzy oil drop model. Notably, TvO and HvO lead us to conclude that O is similarly aligned with T and H. We may therefore speculate that while each protofibril is dominated by the intrinsic hydrophobicity of its component residues, the entire complex (superfibril) forms as a result of interactions between protofibrils which acknowledge the presence of the aqueous environment.

Comparative analysis of protein structures generated using folding simulation software

Similarly to the analysis of the Aβ(15–40) (PDB ID: 2MPZ) [15], we will conduct a comparative study by seeking fragment of the chain whose properties suggest that they may act as amyloid seeds. In other words, the fragments of interest need to be characterized by the following: high value of HvO and negative values of both HvT and TvO. A negative correlation coefficient suggests that the given fragment not only deviates from the reference distribution, but in fact may be regarded as a polar opposite thereof. When such values are accompanied by high RD (above 0.5), we may suspect that the conformation of the given fragment is determined by the intrinsic hydrophobicity of its residues.

As it was discussed formerly [15], to search the origin of the amyloidosis tendency of the polypeptide, the best programs predicting the structure for given amino acid sequence were used to construct alternative structural forms for this polypeptide: I-TASSER [3–5] and ROSETTA [6–9]. These two programs delived 5 alternative structural forms for given sequence (following the CASP project rule [10]). As was said earlier – Robetta delivered 10 models in this case. Additionally, the folding was performed using fuzzy oil drop model (Chapter 2). The FOD model based folding simulation delivers the structures formed by the active participation of water environment. In this study, structures belonging to the output generated by I-Tasser and Robetta are marked by letters I and R respectively. Names of structures generated by FOD start with F. Number following these IDs distinguish the individual models.

Table 10.B.1 lists the FOD coefficients (both values of RD as well as all three correlation coefficients – HvT, TvO and HvO) for each fragment under consideration: 1–6, 7–15, 16–21, 23–28. The study set consists of models obtained using Robetta (R1...R10), I-Tasser (I1...I5) and FOD-based simulations (F1...F5). It also contains results for a representative of native structure of Aβ(1–40) – chain C from 2MVX (as part of superfibril and as an individual unit).

The presented results suggest that structures labeled F1, F2, F3 as well as R1, R2, R3 and R4 are all consistent with the theoretical distribution, implying that the chain is capable of adopting a globular conformation. Table 10.B.1 also highlights forms which exhibit amyloid-like properties – evidenced by negative values of both HvT and TvO, a strongly positive value of HvO, and high values of both RD (T-O-R and T-O-H). All such structures are underscored in the table.

In most cases, however, the RD value for the T-O-H reference model is not particularly high. This may be explained by observing that we are dealing with standalone individual chains – in contrast to chains analyzed as part of an amyloid fibril. Under such conditions the dominant role of intrinsic hydrophobicity is not as evident as could be expected taking into account the structural forms obtained by FOD and Robetta: F1, F2, F3 as well as R1, R2, R3 and R4.

When summarizing the results presented in Table 10.B.1 (see Table 10.B.2 for a compact presentation), it is worth noting that the

Table.1 Fuzzy oil drop parameters for simulated model structures of 2MVX / Aβ(1-40) and their fragments (as denoted by the leftmost column), sorted in order of increasing RD(T-O-R) values for the whole chain. Columns "M" contains model names (which also designate methods used to obtain them): F1…F5 – FOD, I1...I5 – I-Tasser, R1...R10 – Robetta. "CC" stands for chain C from experimentally determined structure of 2MVX treated as part of the complex, while CS – as a standalone structure.

		RD		Correlation Coefficient				RD		Correlation Coefficient				RD		Correlation Coefficient		
FRAGMENT	M	T-O-R	T-O-H	HvT	TvO	HvO	M	T-O-R	T-O-H	HvT	TvO	HvO	M	T-O-R	T-O-H	HvT	TvO	HvO
1-40	F1	0.250	0.216	0.251	0.819	0.378	F2	0.266	0.235	0.246	0.815	0.372	F3	0.279	0.224	0.164	0.799	0.358
1–6		0.448	0.154	-0.319	0.543	-0.329		0.775	0.289	0.831	-0.506	-0.187		0.675	0.329	0.308	-0.041	0.061
07–15		0.264	0.201	0.428	0.831	0.403		0.389	0.310	0.148	0.561	0.359		0.392	0.306	0.365	0.557	0.372
16–21		0.306	0.195	0.188	0.820	0.613		0.448	0.399	0.005	0.552	0.702		0.342	0.292	0.077	0.802	0.593
23–28		0.243	0.234	0.430	0.907	0.676		0.243	0.150	0.166	0.804	0.594		0.321	0.110	-0.214	0.757	0.450
29–40		0.396	0.211	-0.273	0.727	-0.555		0.233	0.130	-0.322	0.846	-0.538		0.245	0.137	-0.417	0.829	-0.538
11–19		0.287	0.196	0.231	0.778	0.652		0.399	0.324	0.115	0.610	0.666		0.436	0.357	0.154	0.576	0.641
1–40	R1	0.434	0.375	0.502	0.578	0.704	R2	0.458	0.438	0.501	0.564	0.694	R3	0.465	0.457	0.459	0.531	0.703
1–6		0.356	0.328	0.674	0.805	0.822		0.203	0.206	0.632	0.892	0.721		0.168	0.139	0.788	0.904	0.880
07–15		0.486	0.874	0.444	0.535	0.945		0.539	0.488	0.215	0.341	0.617		0.717	0.450	-0.121	0.231	0.867
16–21		0.200	0.065	0.829	0.894	0.870		0.240	0.254	0.827	0.855	0.831		0.196	0.282	0.814	0.918	0.836
23–28		0.356	0.058	0.015	0.727	0.527		0.639	0.133	0.072	0.228	0.804		0.604	0.113	0.018	0.253	0.800
29–40		0.482	0.382	0.230	0.781	0.085		0.725	0.573	0.287	0.609	-0.005		0.747	0.595	0.287	0.619	0.033
11–19		0.404	0.277	0.650	0.697	0.929		0.212	0.281	0.786	0.894	0.817		0.139	0.225	0.782	0.957	0.850
1–40	R4	0.465	0.365	0.414	0.565	0.755	R5	0.508	0.506	0.411	0.465	0.791	R6	0.509	0.500	0.555	0.544	0.780
1–6		0.229	0.425	0.832	0.864	0.985		0.378	0.658	0.643	0.656	0.965		0.297	0.478	0.832	0.775	0.958
07–15		0.356	0.349	0.510	0.719	0.739		0.330	0.423	0.505	0.770	0.813		0.493	0.533	0.489	0.595	0.797
16–21		0.112	0.047	0.759	0.934	0.872		0.580	0.139	0.050	0.131	0.748		0.354	0.222	0.852	0.559	0.747
23–28		0.616	0.121	-0.335	0.283	0.479		0.396	0.547	0.340	0.671	0.920		0.592	0.260	0.031	0.251	0.862
29–40		0.770	0.568	-0.050	0.325	-0.119		0.753	0.723	0.206	0.371	0.376		0.749	0.695	0.192	0.267	0.273
11–19		0.117	0.097	0.895	0.975	0.877		0.331	0.159	0.457	0.726	0.869		0.187	0.154	0.856	0.821	0.837
1–40	R7	0.519	0.434	0.412	0.436	0.721	R8	0.532	0.427	0.370	0.515	0.681	I1	0.602	0.409	0.132	0.332	0.516
1–6		0.218	0.101	0.824	0.909	0.804		0.330	0.125	0.235	0.856	0.290		0.421	0.243	0.666	0.472	0.275
07–15		0.618	0.515	-0.075	0.229	0.869		0.746	0.668	-0.103	-0.154	0.913		0.806	0.348	-0.412	0.049	-0.475

16−21		0.255	0.132	0.746	0.815	0.771		0.265	0.155	0.833	0.803	0.840		0.652	0.385	-0.350	-0.591	0.859
23−28		0.422	0.078	0.231	0.509	0.857		0.727	0.190	-0.119	0.010	0.806		0.814	0.253	-0.866	-0.609	0.291
29−40		0.790	0.684	0.231	0.345	0.149		0.767	0.687	0.289	0.323	0.145		0.792	0.744	0.366	0.521	0.287
11−19		0.136	0.110	0.724	0.937	0.828		0.145	0.120	0.762	0.923	0.824		0.492	0.222	-0.173	0.363	0.526
1−40	CC	0.608	0.620	0.459	0.665	0.784	I2	0.623	0.508	0.147	0.218	0.540	F4	0.629	0.368	-0.157	0.312	0.238
1−6		0.332	0.207	0.621	0.739	0.770		0.352	0.157	-0.317	0.680	0.322		0.506	0.111	-0.406	0.593	-0.283
07−15		0.857	0.857	-0.113	-0.198	0.850		0.762	0.641	0.222	-0.250	0.433		0.623	0.499	0.152	0.727	0.531
16−21		0.407	0.282	0.485	0.622	0.946		0.677	0.602	-0.593	-0.905	0.792		0.581	0.358	-0.020	0.196	0.726
23−28		0.533	0.461	0.065	0.039	0.995		0.648	0.108	-0.643	-0.153	0.403		0.680	0.293	-0.958	-0.342	0.417
29−40		0.853	0.648	0.157	0.301	-0.005		0.758	0.617	0.140	0.344	-0.285		0.548	0.386	-0.260	0.607	-0.672
11−19		0.565	0.484	0.285	0.348	0.929		0.564	0.482	-0.110	-0.182	0.776		0.371	0.215	0.156	0.631	0.697
1−40	F5	0.631	0.433	-0.122	0.350	0.285	I3	0.634	0.372	0.134	0.401	0.552	CS	0.636	0.562	0.295	0.363	0.616
1−6		0.466	0.329	-0.101	0.530	0.251		0.576	0.325	-0.162	0.185	0.622		0.682	0.370	0.747	0.025	0.233
07−15		0.756	0.454	0.358	0.356	0.295		0.859	0.600	-0.325	-0.283	0.173		0.864	0.783	-0.062	-0.405	0.606
16−21		0.542	0.324	-0.767	0.145	0.397		0.499	0.162	0.330	0.492	0.876		0.456	0.171	0.199	0.407	0.844
23−28		0.821	0.443	-0.920	-0.903	0.754		0.790	0.122	-0.797	-0.148	0.015		0.421	0.230	0.494	0.508	0.994
29−40		0.659	0.506	0.040	0.587	-0.514		0.884	0.723	0.204	0.626	-0.139		0.797	0.690	0.159	0.611	-0.022
11−19		0.473	0.217	-0.144	0.618	0.211		0.407	0.177	0.169	0.624	0.627		0.568	0.362	0.213	0.258	0.809
1−40	I4	0.642	0.459	0.277	0.333	0.491	R9	0.650	0.601	0.372	0.413	0.757	I5	0.702	0.521	0.266	0.208	0.533
1−6		0.419	0.531	0.805	0.532	0.847		0.602	0.555	0.357	-0.157	0.684		0.740	0.566	0.403	0.381	0.917
07−15		0.886	0.502	-0.099	-0.447	0.200		0.685	0.892	0.317	0.155	0.909		0.893	0.673	-0.079	-0.836	0.183
16−21		0.722	0.467	0.124	-0.460	0.615		0.723	0.297	-0.393	-0.394	0.798		0.662	0.510	-0.006	-0.459	0.693
23−28		0.808	0.299	-0.760	-0.454	0.665		0.459	0.312	0.275	0.440	0.914		0.837	0.222	-0.844	-0.452	0.143
29−40		0.502	0.357	0.185	0.618	-0.392		0.737	0.644	0.119	0.454	0.169		0.728	0.462	0.473	0.410	-0.504
11−19		0.446	0.231	0.254	0.444	0.673		0.509	0.320	0.087	0.217	0.827		0.542	0.340	0.038	0.199	0.711
1−40	R10	0.718	0.612	0.142	0.025	0.680												
1−6		0.669	0.467	0.781	0.062	0.615												
07−15		0.782	0.630	-0.070	-0.390	0.701												
16−21		0.527	0.135	0.027	0.129	0.802												
23−28		0.269	0.080	0.676	0.810	0.965												
29−40		0.724	0.822	0.587	0.422	0.693												
11−19		0.515	0.160	-0.012	0.244	0.886												

Table 10.B.2 Status of suspected amyloid seeds in simulated model structures of Aβ(1−40), compared with experimental observations (2MVX structure). Numbers in the table correspond to names of models produced by each software package. The rightmost column presents the experimentally determined status for native chain C as part of the complex as a whole as well as its individual chains when treated as an individual unit.

Fragment	FOD	I-Tasser	Robetta	Aβ(1−40) (2MVX)
Accordant fragments (RD < 0.5)				
1−40	F1,F2,F3		R1,R2,R3,R4	
1−6	F1,F5	I1,I2,I4	R1,R2,R3,R4,R5,R6, R7,R8	Comp.
7−15	F1,F2,F3		R1,R4,R5,R6	
16−21	F1,F2,F3	I3	R1,R2,R3,R4,R6, R7,R8	Comp., Indiv.
23−28	F1,F2,F3		R1,R5,R7,R9,R10	Indiv.
29−40	F1,F2,F3		R1	
Discordant fragments (RD ≥ 0.5)				
1−40	F4,F5	I1,I2,I3,I4,I5	R5,R6,R7,R8,R9,R10	Comp., Indiv.
1−6	F2,F3,F4	I3,I5	R9,R10	Indiv.
7−15	F4,F5	I1,I2,I3,I4,I5	R2,R3,R7,R8,R9,R10	Comp., Indiv.
16−21	F4,F5	I1,I2,I4,I5	R5,R9,R10	
23−28	F4,F5	I1,I2,I3,I4,I5	R2,R3,R4,R6,R8	Comp.
29−40	F4,F5	I1,I2,I3,I4,I5	R2,R3,R4,R5,R6,R7, R8,R9,R10	Comp., Indiv.

fragments at 7−15, 16−21 and 22−27, 23−28 frequently deviate from the theoretical monocentric distribution in favor of experimentally determined properties (except for the 16−21, where such conditions are not observed).

When discussing the status of each fragment it should be noted that the evaluation criteria are supplied by the FOD model.

The 3D structures of generated models are shown in Fig. 10.B.4 (F1…F5, I1…F5, R1 … R5) and Fig. 10.B.5 (R6 … R10). Hydrophobicity profiles of selected structures accordant and discordant with the fuzzy oil drop model are shown in Fig. 10.B.6 and Fig. 10.B.7 respectively.

Analysis of results shown in Table 10.B.1 and Table 10.B.2 reveals − somewhat surprisingly − that the models produced by Robetta are largely consistent with the monocentric distribution of hydrophobicity. While the Robetta algorithm does not directly account for internalization of hydrophobic residues, it nevertheless proves that the presented chain may,

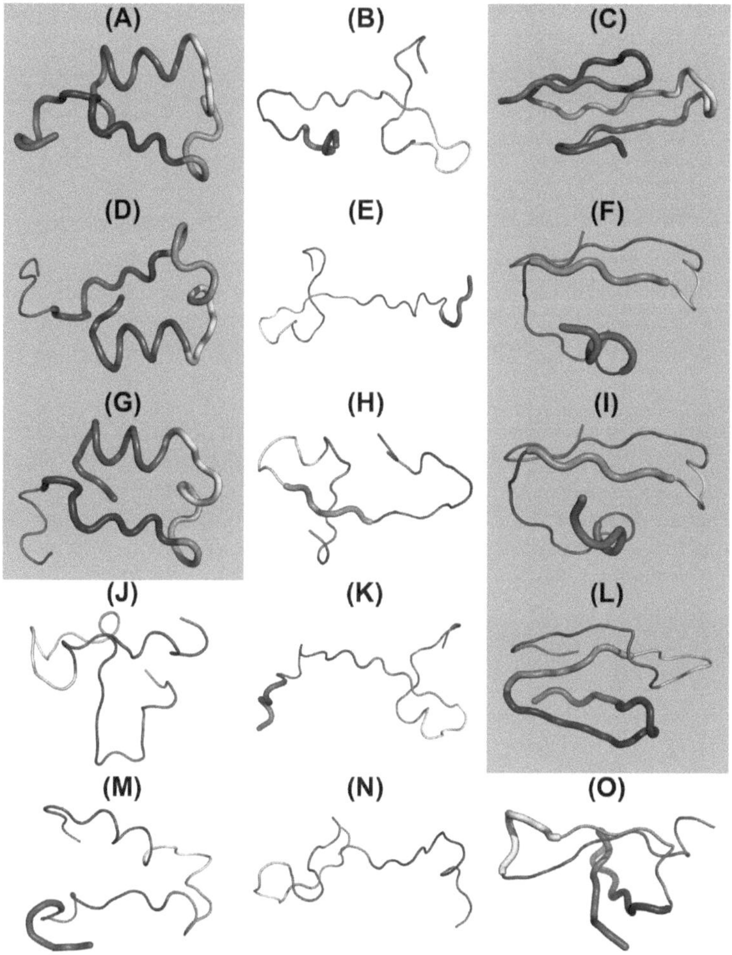

Fig. 10.B.4 3D presentation of simulated model structures of Aβ(1–40) (2MVX), part 1. A, D, G, J, M– F1...F5. B, E, H, K, N – I1–I5. C, F, I, L, O – R1 ... R5. Fragments analyzed in this study are highlighted by colors: red – 1–6, blue – 7–15, green – 16–21, yellow – 23–28. Rest of each chain (29–40) is gray. Gray background denotes RD (T-O-R) < 0.5 for the whole chain (1–40). Status of each fragment of the structures is given by the backbone trace style: thick – accordant (RD < 0.5), thin – discordant (RD ≥ 0.5).

under certain circumstances, adopt a globular conformation. It is interesting to speculate why such phenomena are not observed *in vivo*.

In summary, it is worth noting that Robetta generated four models whose RD is lower than 0.5 (indicating the presence of a centralized

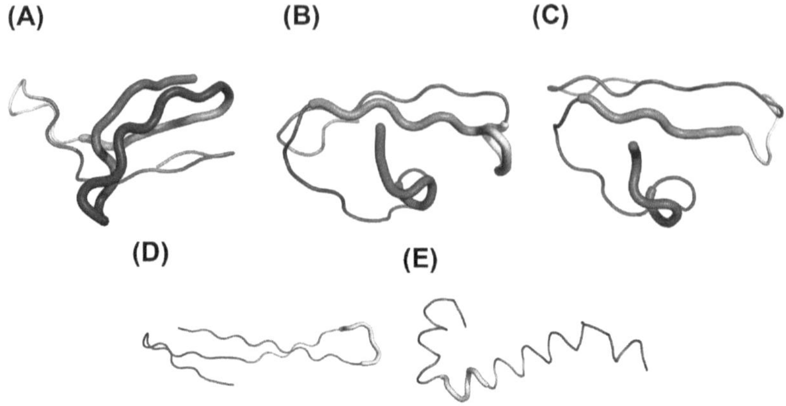

Fig. 10.B.5 3D presentation of simulated model structures of Aβ(1–40) (2MVX), part 2. A, B, C, D, E R6 … R10. Fragments analyzed in this study are highlighted by colors: red – 1–6, blue – 7–15, green – 16–21, yellow – 23–28. Rest of each chain (29–40) is gray. Gray background denotes RD (T-O-R) < 0.5 for the whole (1–40) chain. Status of each fragment of the structures is given by the backbone trace style: thick – accordant (RD $<$ 0.5), thin – discordant (RD $\geq$ 0.5).

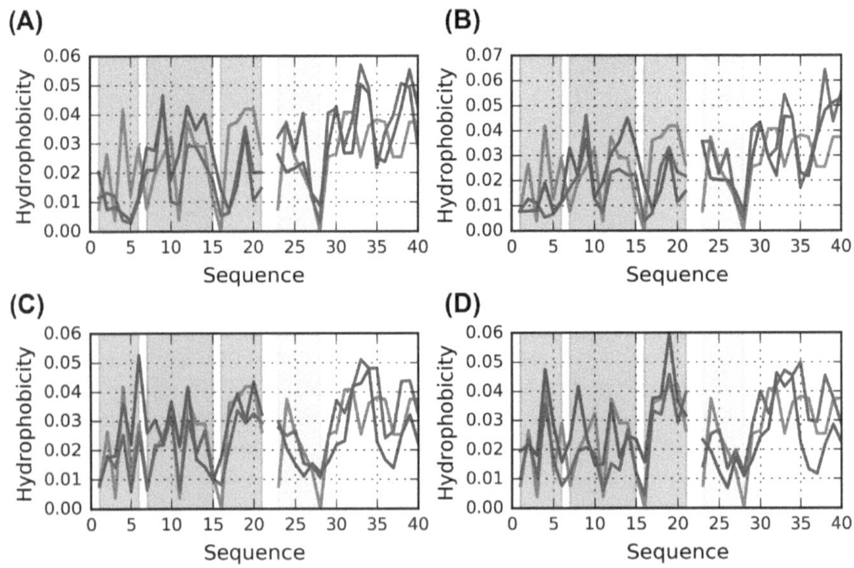

Fig. 10.B.6 Theoretical (T, blue), observed (O, red) and intrinsic (H, green) hydrophobicity distribution profiles for selected accordant simulated model structures of Aβ(1–40) (2MVX). (A) F1; (B) F2; (C) R1; (D) R2. Colored backgrounds highlight fragments analyzed in this study: red – 1–6, blue – 7–15, green – 16–21, yellow – 23–28.

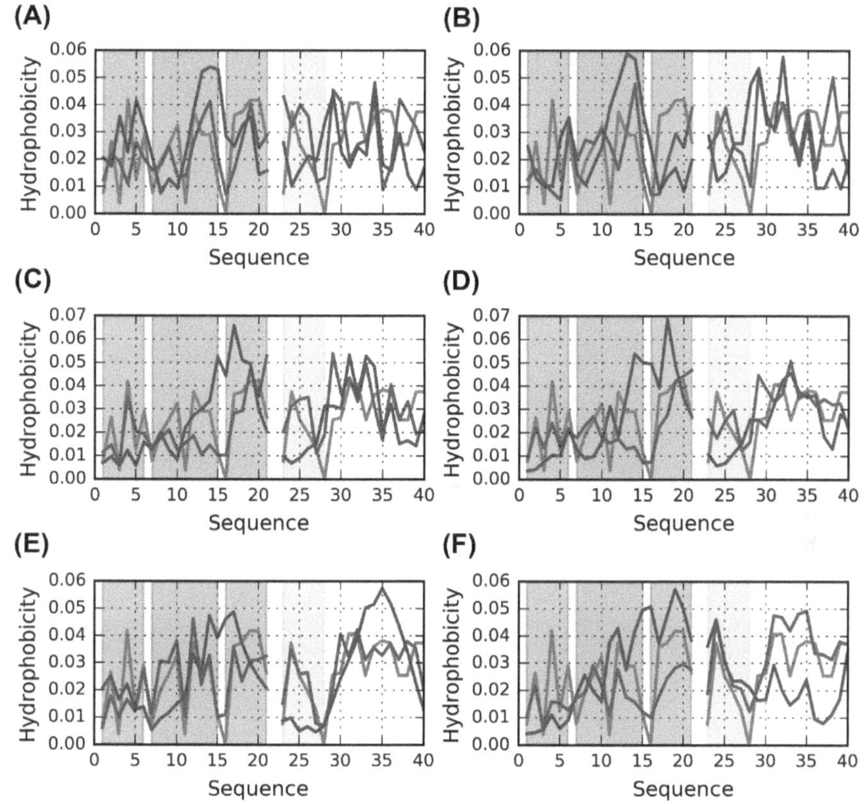

Fig. 10.B.7 Theoretical (T, blue), observed (O, red) and intrinsic (H, green) hydrophobicity distribution profiles for selected discordant simulated model structures of Aβ(1−40) (2MVX). Colored backgrounds highlight fragments analyzed in this study: red − 1−6, blue − 7−15, green − 16−21, yellow − 23−28.

hydrophobic core). In contrast, I-Tasser produced no such models. While I-Tasser models are generally more reminiscent of amyloid forms (Fig. 10.B.4), the greatest deviation from the monocentric core pattern is observed for one of the presented Robetta models. Regarding FOD, the computed models promote internalization of hydrophobic residues but are nevertheless quite divergent from globular forms [17,18].

References

[1] Schütz AK, Vagt T, Huber M, Ovchinnikova OY, Cadalbert R, Wall J, Meier BH. Atomic-resolution three-dimensional structure of amyloid β fibrils bearing the Osaka mutation. Angewandte Chemie International Edition 2014;54(1):331−5. https://doi.org/10.1002/anie.201408598.

[2] Tycko R. Molecular structure of aggregated amyloid-β: insights from solid-state nuclear magnetic resonance. Cold Spring Harbor Perspectives in Medicine 2016;6(8): a024083. https://doi.org/10.1101/cshperspect.a024083.
[3] Zhang Y. I-tasser server for protein 3D structure prediction. BMC Bioinformatics 2008;9(1):40. https://doi.org/10.1186/1471-2105-9-40.
[4] Roy A, Kucukural A, Zhang Y. I-tasser: a unified platform for automated protein structure and function prediction. Nature Protocols 2010;5(4):725–38. https://doi.org/10.1038/nprot.2010.5.
[5] Yang J, Yan R, Roy A, Xu D, Poisson J, Zhang Y. The I-tasser Suite: protein structure and function prediction. Nature Methods 2015;12(1):7–8. https://doi.org/10.1038/nmeth.3213.
[6] https://zhanglab.ccmb.med.umich.edu/I-TASSER.
[7] http://robetta.org.
[8] Kim DE, Chivian D, Baker D. Protein structure prediction and analysis using the Robetta server. Nucleic Acids Research 2004;32(Web Server):W526–31. https://doi.org/10.1093/nar/gkh468.
[9] http://predictioncenter.org.
[10] Moult J, Fidelis K, Kryshtafovych A, Schwede T, Tramontano A. Critical assessment of methods of protein structure prediction (CASP) — round x. Proteins: Structure, Function, and Bioinformatics 2013;82:1–6. https://doi.org/10.1002/prot.24452.
[11] Konieczny L, Brylinski M, Roterman I. Gauss-function-based model of hydrophobicity density in proteins. Silico Biology 2006;6(1–2):15–22.
[12] Kalinowska B, Banach M, Konieczny L, Roterman I. Application of divergence entropy to characterize the structure of the hydrophobic core in DNA interacting proteins. Entropy 2015;17(3):1477–507. https://doi.org/10.3390/e17031477.
[13] Kullback S, Leibler RA. On information and sufficiency. The Annals of Mathematical Statistics 1951;22(1):79–86. https://doi.org/10.1214/aoms/1177729694.
[14] Dułak D, Gadzała M, Banach M, Ptak M, Wiśniowski Z, Konieczny L, Roterman I. Filamentous aggregates of tau proteins fulfil standard amyloid criteria provided by the fuzzy oil drop (FOD) model. International Journal of Molecular Sciences 2018;19(10): 2910. https://doi.org/10.3390/ijms19102910.
[15] Dułak D, Banach M, Gadzała M, Konieczny L, Roterman I. Structural analysis of the Aβ(15–40) amyloid fibril based on hydrophobicity distribution. Acta Biochimica Polonica 2018. https://doi.org/10.18388/abp.2018_2647.
[16] Laskowski RA, Jabłońska J, Pravda L, Vařeková RS, Thornton JM. PDBsum: structural summaries of PDB entries. Protein Science 2017;27(1):129–34. https://doi.org/10.1002/pro.3289.
[17] Schmit JD, Ghosh K, Dill K. What drives amyloid molecules to assemble into oligomers and fibrils? Biophysical Journal 2011;100(2):450–8. https://doi.org/10.1016/j.bpj.2010.11.041.
[18] Banach M, Konieczny L, Roterman I. Fuzzy oil drop model application—from globular proteins to amyloids. Computational Methods to Study the Structure and Dynamics of Biomolecules and Biomolecular Processes 2018:639–58. https://doi.org/10.1007/978-3-319-95843-9_19.

CHAPTER 10.C

Specificity of amino acid sequence and its role in secondary and supersecondary structure generation

Mateusz Banach[1], **Irena Roterman**[1]
[1]Department of Bioinformatics and Telemedicine, Jagiellonian University—Medical College, Krakow, Poland

Contents

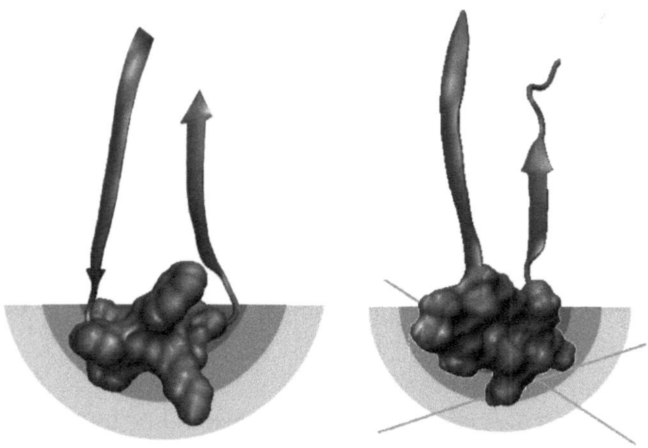

Schematic presentation of loops (7 aa) linking two β-strands (fragments of β-sheets). The left one — helical form represents fuzzy oil drop distribution of

From Globular Proteins to Amyloids
ISBN: 978-0-08-102981-7
https://doi.org/10.1016/B978-0-08-102981-7.00018-X

hydrophobicity. The right one not able to generate the helical form represents the status recognized as amyloid seed in amyloids discussed in this work.

The role of amino acids sequence — in particular 7aa fragment linking β-strands belonging to β-sheets can be easily visualized discussing examples of proteins generating the amyloid-like structures with the one representing the status accordant with the fuzzy oil drop model.

Short fragments of polypeptide playing the ole of linkers between β-strands belonging to β-sheets

Three polypeptides: Aβ(15—40) — amyloid form (PDB ID: 2MPZ, "Iowa" mutant D23N) [1], chain C Aβ(16—40) "packaged" with a synthetic protein in the complex with protein (PDB ID: 2OTK) [2] and C domain of light chain of IgG (PDB ID: 7FAB) [3] are taken as examples for the following analysis. These three structures contain β-strands belonging to β-sheets. These β-strands are linked by 7 aa fragment. The difference between selected proteins is their status as whole.

The C domain of light chain of IgG represents the hydrophobicity distribution accordant with fuzzy oil drop model. The status of 7 aa polypeptide chain fragments linking β-strands participating in β-sheet is examined. The fragments 22—28 of amyloid polypeptide chains (as observed in 2MPZ and 2OTK) and fragment 118—124 are compared to visualize the role and specificity of the sequence.

A proposed experiment involves modifying the 22—28 sequence to match the 118—124 fragment in the light chain of IgG. The amphipathic properties of the helical linker would likely mediate entropically advantageous contact with the aqueous solvent and therefore affect the structure as a whole. The 118—124 fragment is clearly predisposed toward a helical conformation, as listed in the database of chameleon sequences [4] where values greater than 1 indicate structural affinity, and the helical form scores more than 3 [4]. The amphipathic helix may be regarded as the an alternative to a single dominant maximum which introduces strong discordance between O and T.

It is also interesting to note the status of the C-terminal fragment, which remains highly accordant in the immunoglobulin but is excessively hydrophobic in the Aβ chain, despite remaining in contact with water (note the low values of O).

The β-hairpin is a common structural motif, found in many proteins. The β-hairpin discussed here differs in respect to proper β-hairpin

Table 10.C.1 Fuzzy oil drop parameters for 7FAB, 2OTK and 2MPZ, along with fragments linking β-hairpin folds, compared to the corresponding linkers in the Aβ(1–42) amyloid. Values listed in boldface represent linkers between separate β strands.

		RD		Correlation coefficient		
Structure	**Fragment**	**T-O-R**	**T-O-H**	**HvT**	**TvO**	**HvO**
7FAB (chain L)	104–204	0.320	0.237	0.588	0.809	0.825
	110–136	0.232	0.308	0.670	0.893	0.863
	118–124	**0.374**	**0.313**	**0.076**	**0.619**	**0.826**
	167–194	0.291	0.266	0.606	0.790	0.809
	178–184	**0.361**	**0.240**	**0.830**	**0.617**	**0.784**
2OTK (chain C)	16–40	0.592	0.290	0.279	0.494	0.603
	22–28	**0.613**	**0.356**	**−0.306**	**−0.018**	**0.866**
2MPZ (chain S)	15–40	0.627	0.467	0.355	0.351	0.616
	22–28	**0.637**	**0.503**	**−0.337**	**−0.320**	**0.938**

containing 7 instead of 4 residues building the turn. Thus, our analysis of the specificity of the 22–28 fragment is based on comparing it with an accordant β-hairpin (i.e. a structure consistent with the theoretical distribution of hydrophobicity). Table 10.C.1 provides a comparative overviews of all presented structural motifs, including the β-hairpin fragment of Aβ(16–40) in complex with two other chains which serve as permanent chaperones (PDB ID: 2OTK).

All presented fragments are of equal length (7 aa). Taken as a whole, the C domain of the IgG light chain follows a micelle-like distribution, with the linker exhibiting low RD and balanced values of FOD correlation coefficients. Radically different conditions are observed in the Aβ(15–42) chain, for reasons illustrated in Fig. 10.C.1.

The observed differences seem to be caused by the fact that in the case of IgG the beta folds are linked by a helical fragment, which is largely accordant with the theoretical distribution (as evidenced by its low RD) thanks to its amphipatic character (see Fig. 10.C.1 and Table 10.C.1). In all presented cases, this linker (with a length of 7 aa) connects two separate β folds which belong to different β sheets. This is particularly evident in the case of Aβ(15–40) (Fig. 10.C.2). Similarly, in the C domain of the IgG light chain the aforementioned helix (178–182) links two distinct β sheets. The intrinsic hydrophobicity which is able to generate the amphipatic helix

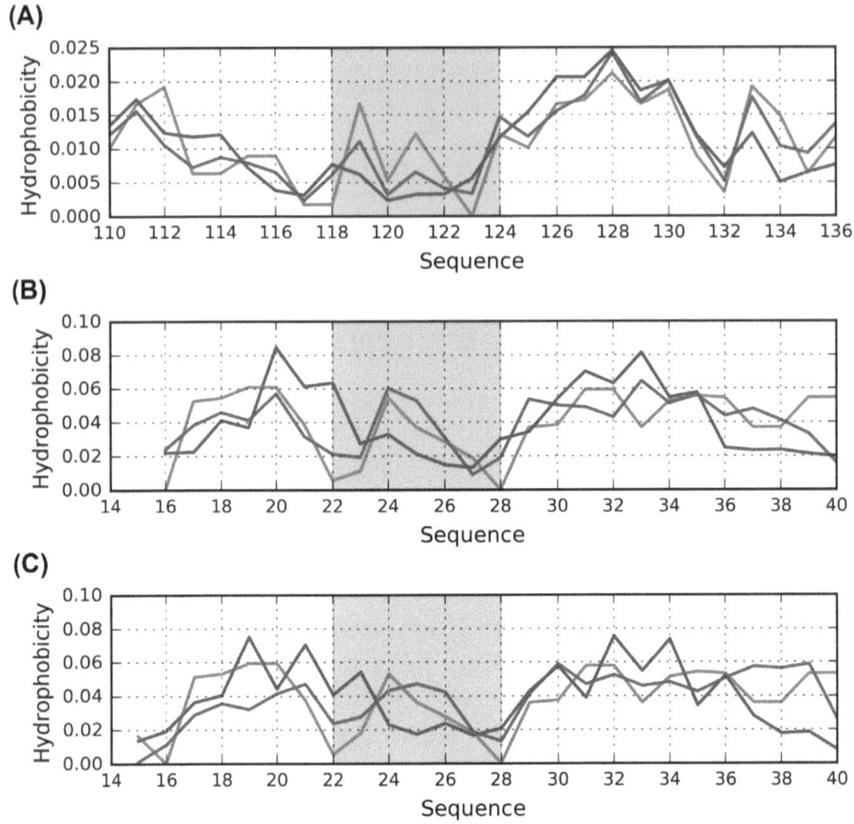

Fig. 10.C.1 Theoretical (T, blue), observed (O, red) and intrinsic (H, green) hydrophobicity distribution profiles for selected structures containing the β-hairpin motif. (A) C domain of IgG light chain – fragment 104–204 of chain L of 7FAB (view limited to residues 110–136). (B) Aβ(16–40) (chain C of 2OTK). (C) Aβ(15–40) (chain S of 2MPZ). The discussed β-hairpin folds present in the structures are marked by the red background.

(zigzag pattern) is present in immunoglobulin domain. This is not the case in Aβ(15–40).

The visual analysis is possible due to the presentation of 3D structures of amyloid forms (Fig. 10.C.3). Two examples are shown in forms of β-strands linked by accordant helices (Fig. 10.C.4).

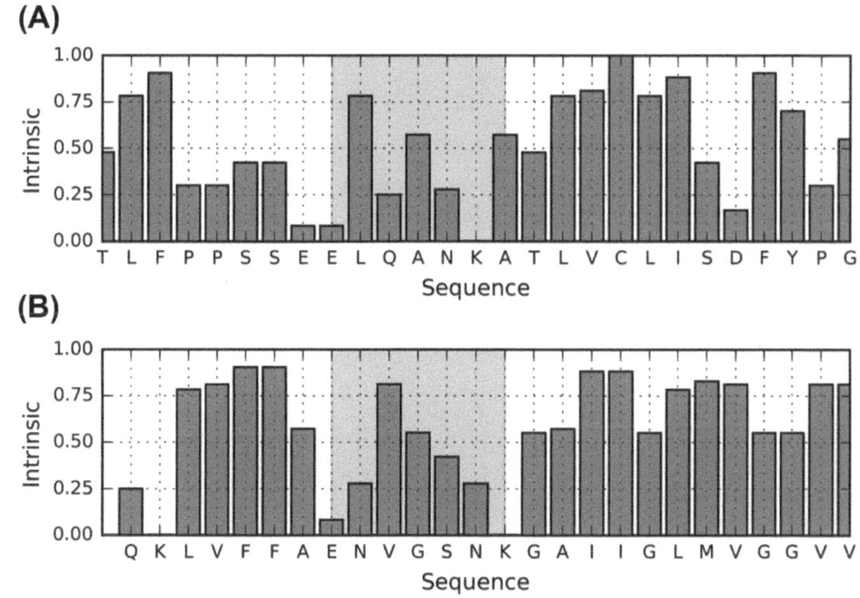

Fig. 10.C.2 Intrinsic hydrophobicity distribution profiles for selected structures containing the β-hairpin motif. (A) C domain of IgG light chain – fragment 104–204 of chain L of 7FAB (view limited to residues 110–136). (B) Aβ(15–40) (chain S of 2MPZ). The discussed turns in β-hairpin folds present in the structures are marked by the red background. A zigzag form of intrinsic hydrophobicity for helical fragment be seen. This form produces the amphipatic helices.

Short peptides

Experimental studies of short peptide sequences provide support for the hypotheses presented in this work. Sample sequences are described in Refs. [5,6].

It appears that the presented sequence fulfills the conditions for linear propagation criteria. Short peptides typically do not have a tertiary conformation and therefore cannot produce a spherical micelle. Thus, for structures such as the one shown in Fig. 10.C.5, a ribbonlike micelle remains the only option.

The specifics of protein-water interaction remain an open issue. If a sufficiently long hydrophobic band appears in the environment, the question of its influence upon the solvent touches upon the central aspect of the mechanism which guides protein folding (as well as misfolding). It is also useful to consider the reaction of water to the presence of various types of protein surfaces.

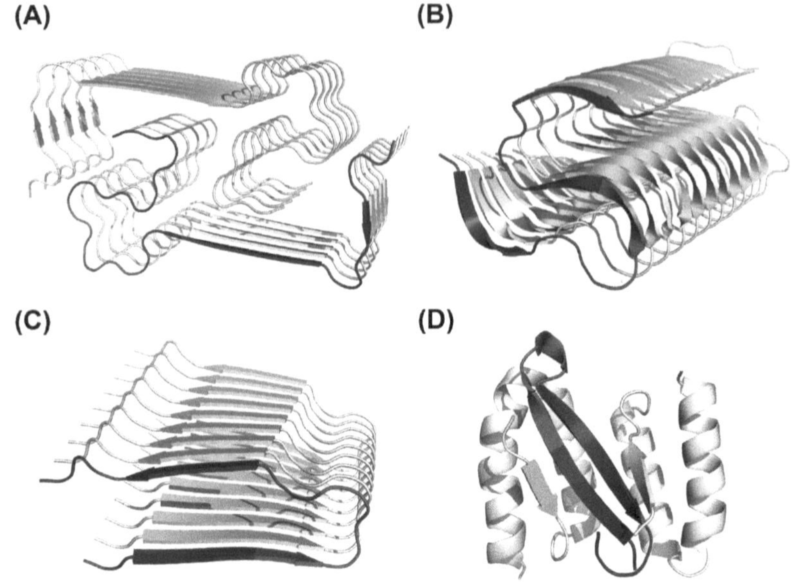

Fig. 10.C.3 3D presentation of location of 22–28 fragment in selected Aβ fragments. (A) Aβ(1–40) (2MVX). (B) Aβ(11–42) (2MXU). (C) Aβ(15–40) (2MPZ – protofibril). (D) Aβ(16–40) (2OTK). In 2MVX, 2MXU and 2MPZ, an example chain in the fibril is shown in blue, with 22–28 fragment highlighted in red. In 2OTK, the Aβ(16–40) is presented in complex with two external chains which act as its permanent chaperones.

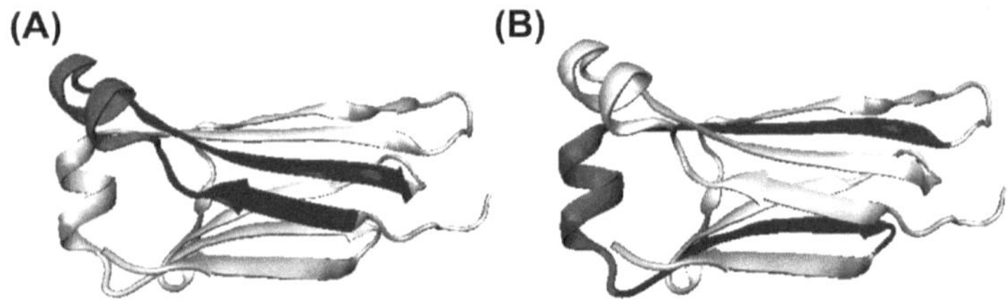

Fig. 10.C.4 3D presentation of C domain of light chain of IgG (7FAB, chain L). (A) location of β-hairpin motif (red, residues 118–124) in the 110–136 fragment (blue). (B) two β-strands (sand, residues 168–193) linked by helical turn (red, 178–184).

Alternative structures solve the problem of minimization of the hydrophobic area on the surface generating the ribbon-like structural forms

The goal of the analysis presented in this chapter is to reveal the potential structural variability of the Aβ(1–40) sequence. It turns out that – much like the previously discussed cases, i.e. Aβ(15–40) [7], and the tau

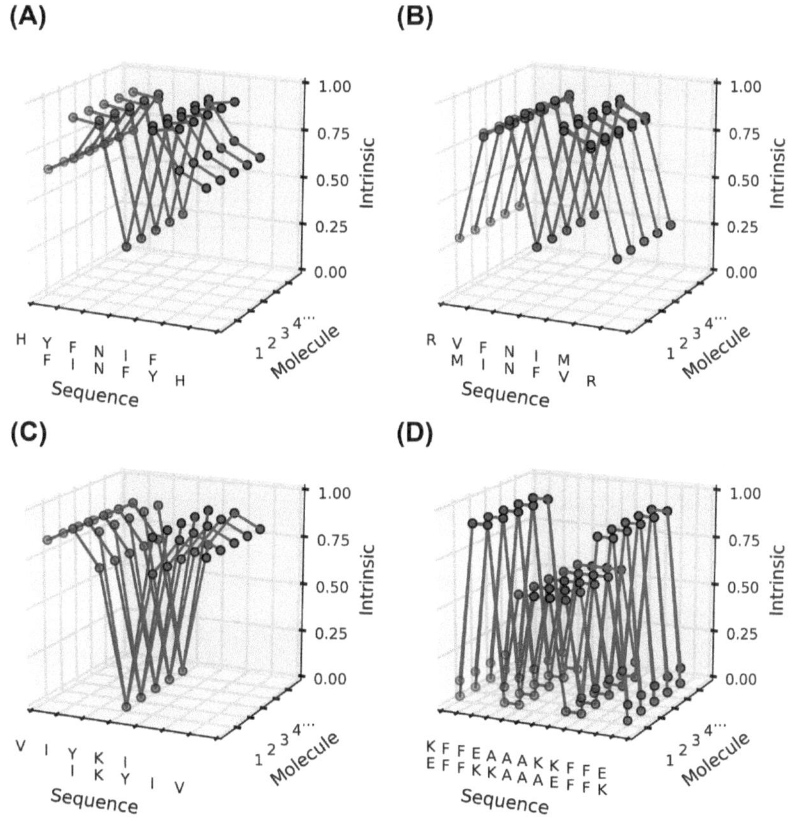

Fig. 10.C.5 Intrinsic hydrophobicity profiles for the anti-parallel orientation of β-strands in peptides described in Ref. [5]: blue – top sequence, red – bottom sequence. (A) HYFNIF; (B) RVFNIM; (C) VIYKI, (D) KFFEAAAKKFFE.

amyloid [8] – such chains are possibly able of producing globular structures. This observation is supported by results of simulations carried out using specialized folding software (I-Tasser and Robetta) (Chapter 10.B).

The capability to produce a monocentric hydrophobic core suggest that such structures may emerge under real-world conditions, even in the absence of a guiding factor (i.e. in the case of Robetta models). Consequently, we may speculate that the specific structural pattern adopted by a polypeptide chain depends on its environment and that in some cases environmental conditions may favor creation of structures dominated by intrinsic hydrophobicity.

In [9] the authors postulate that even minute structural changes in the solvent may shift the balance toward oligomerization and fibrillarization.

This suggestion is quantitatively supported by simulations carried out using the fuzzy oil drop model, where deviations from the theoretical distribution of hydrophobicity – both local and global – may result in formation of elongated fibrillar forms. The comparison of globular molecules versus the amyloid is show also in Chapter 9.

Given our objective – i.e. devising a way to identify amyloid seeds – modifying the 22–28 fragment of Aβ in order to produce a helical fold is a useful test of the presented theory.

References

[1] Sgourakis NG, Yau W-M, Qiang W. Modeling an in-register, parallel "Iowa" Aβ fibril structure using solid-state NMR data from labeled samples with rosetta. Structure 2015; 23(1):216–27. https://doi.org/10.1016/j.str.2014.10.022.

[2] Hoyer W, Grönwall C, Jonsson A, Ståhl S, Härd T. Stabilization of a β-hairpin in monomeric Alzheimer's amyloid-β peptide inhibits amyloid formation. Proceedings of the National Academy of Sciences 2008;105(13):5099–104. https://doi.org/10.1073/pnas.0711731105.

[3] Saul FA, Poljak RJ. Crystal structure of human immunoglobulin fragment Fab new refined at 2.0 Å resolution. Proteins: Structure, Function, and Genetics 1992;14(3): 363–71. https://doi.org/10.1002/prot.340140305.

[4] Ghozlane A, Joseph AP, Bornot A, de Brevern AG. Analysis of protein chameleon sequence characteristics. Bioinformation 2009;3(9):367–9. PMCID:PMC2732029.

[5] Al-Garawi ZS, Morris KL, Marshall KE, Eichler J, Serpell LC. The diversity and utility of amyloid fibrils formed by short amyloidogenic peptides. Interface Focus 2017;7(6): 20170027. https://doi.org/10.1098/rsfs.2017.0027.

[6] Marshall KE, Vadukul DM, Dahal L, Theisen A, Fowler MW, Al-Hilaly Y, Serpell LC. A critical role for the self-assembly of Amyloid-β1-42 in neurodegeneration. Scientific Reports 2016;6(1). https://doi.org/10.1038/srep30182.

[7] Dułak D, Banach M, Gadzała M, Konieczny L, Roterman I. Structural analysis of the Aβ(15–40) amyloid fibril based on hydrophobicity distribution. Acta Biochimica Polonica 2018. https://doi.org/10.18388/abp.2018_2647.

[8] Dułak D, Gadzała M, Banach M, Ptak M, Wiśniowski Z, Konieczny L, Roterman I. Filamentous aggregates of tau proteins fulfil standard amyloid criteria provided by the fuzzy oil drop (FOD) model. International Journal of Molecular Sciences 2018; 19(10):2910. https://doi.org/10.3390/ijms19102910.

CHAPTER 11

Anti-amyloid drug design

Mateusz Banach[1], Leszek Konieczny[2], Irena Roterman[1]
[1]Department of Bioinformatics and Telemedicine, Jagiellonian University – Medical College, Krakow, Poland
[2]Chair of Medical Biochemistry, Jagiellonian University – Medical College, Krakow, Poland

Contents

Conceptual visualization of drug design via complexation of amphipathic helices(in red) compatible with the distribution of hydrophobicity in the fibril and exposing a hydrophilic layer, which facilitates interaction with water. This idea in based on the analysis of stop signals in proteins with linear propagation present in their structure.

If we support the conclusions which arise from applying the fuzzy oil drop model to amyloid structures and proteins which contain solenoid fragments, the process of designing drugs capable of arresting linear propagation (which leads to unrestricted growth of the molecule) should begin with the analysis of ways in which this kind of propagation is prevented in biological proteins containing amyloid-like structures.

The solenoid is a supersecondary structure which results from linear alignment of polypeptide chain fragments, much like in amyloids. Proteins in which such structures appear—mostly lyases and antifreeze proteins—provide "stop fragments" (or "caps") which prevent unrestricted propagation of solenoids. Following this observation, we performed an analysis of both groups of proteins, focusing on their caps. As it turns out, these caps may adopt various conformations—helices (in most cases), random coils and even short β-strands. Additionally, in a handful of cases, no obvious "caps" can be identified; instead the terminal fragment of the solenoid itself exhibits a distribution of hydrophobicity consistent with the micellar form.

From Globular Proteins to Amyloids
ISBN: 978-0-08-102981-7
https://doi.org/10.1016/B978-0-08-102981-7.00019-1

Such fragments can prevent further complexation without the need of a dedicated "cap".

When analyzing the role of fragments whose purpose is to prevent propagation of complexation of additional molecules (resulting in elongation of the solenoid), it is apparent that the fragment should, on the one hand, prevent access to the ordered portion of the solenoid, while on the other hand enabling contact with water and facilitating solubility.

Table 11.1 lists the structural properties of proteins where such "stopper" fragments have been found. Figs. 11.1–11.19 present hydrophobicity profiles and 3D visualizations of these structures.

Analysis of data listed in Table 11.1 indicates that the "stopper" is typically a short helix, although in some cases it may adopt the conformation of a random coil or even a short β-strand. The latter two structures must enter into a specific relation with the remainder of the solenoid (Fig. 11.1). From the point of view of drug design, the helical conformation is preferred. The "stop" fragment should meet several conditions: (1) It should exhibit affinity for the tip of the solenoid, i.e. its conformation should be compatible with that of the outermost solenoid loop (or the outermost peptide in an amyloid fibril); (2) Its outer surface should not repel water. A helix—particularly an amphipathic one—can fulfill both requirements simultaneously. In order to achieve this, the helix should be designed in such a way as to remain compatible with the distribution of hydrophobicity presented by the solenoid (or peptide which needs to be locked out), and to expose polar fragments capable of mediating contact with the aqueous environment. Examples of such short helices which meet the stated conditions and have been designed to math specific amyloid constructs are discussed in Refs. [18,19].

Designing β-strand which possess the required characteristics and are able to arrest propagation of amyloids is much harder due to requirements associated with spatial alignment with the amyloid. By its nature, a β-strand is capable of forming hydrogen bonds in two opposite directions, thus permitting complexation with other β-strands. The alignment must be such as to prevent the fold from attracting additional folds when the given fragment is bound to the solenoid. As illustrated, the orientation of such folds is tricky and complicated, and designing them poses substantial challenges. In most cases, preventing complexation of additional folds calls for another fragment, which must be oriented at an angle with respect to the surface of the amyloid (see Fig. 11.2 for an example). This unusual alignment introduces a special

Table 11.1 Values of fuzzy oil drop parameters calculated for selected structures (structural units for which the 3D Gass function was defined) and "stop" fragments found within their sequences. Asterisks (*) indicate that the β-strand treated as stop fragment is an integral part of the solenoid First line describes the protein (identified by "chain")—for this unit the 3D Gasuss function was calculated; the second line (or more) describes status of polypeptide chain fragment treated as "stopper".

Protein	Fragment		RD		Correlation coefficient			Ref.
	Structure	Residues	T-O-R	T-O-H	HvT	TvO	HvO	
2ZU0		CHAIN	0.645	0.591	0.233	0.389	0.749	[1]
	HELIX	94–104	0.259	0.372	0.446	0.818	0.745	
Antifreeze proteins								
1L0S		CHAIN	0.526	0.418	0.399	0.470	0.752	[2]
	BETA	12–15	0.217	0.140	-0.201	0.903	0.019	
	BETA	72–80	0.443	0.671	0.214	0.548	0.809	
1M8N		CHAIN	0.656	0.603	0.248	0.361	0.784	[3]
	BETA*	2–15	0.395	0.472	0.268	0.636	0.723	
	BETA*	11–15	0.312	0.161	0.353	0.873	0.356	
	RC	106–112	0.567	0.316	0.125	0.327	0.937	
3VN3		CHAIN	0.714	0.613	0.309	0.428	0.685	[4]
	BETA	48–61	0.653	0.507	-0.050	0.397	0.160	
	HELIX	100–109	0.268	0.146	0.166	0.840	0.306	
3P4G		CHAIN	0.753	0.690	0.216	0.384	0.728	[5]
	BETA	23–35	0.566	0.554	0.199	0.534	0.649	
	HELIX	285–302	0.216	0.137	0.660	0.846	0.850	
1Z2F		CHAIN	0.671	0.711	0.186	0.377	0.625	[6]
	BETA	1–7	0.497	0.682	0.295	0.431	0.898	
	RC	102–117	0.476	0.486	0.472	0.622	0.616	
1N4I		CHAIN	0.480	0.398	0.362	0.556	0.787	[7]
	RC	1–9	0.425	0.401	0.100	0.683	0.701	
	RC	71–78	0.313	0.458	0.218	0.752	0.673	
3WP9		CHAIN	0.658	0.576	0.282	0.450	0.679	[8]
	HELIX	40–54	0.671	0.640	0.243	0.069	0.618	
	MIXED	59–70	0.378	0.260	0.327	0.705	0.545	
Lyases								
1BN8		CHAIN	0.684	0.559	0.194	0.346	0.750	[9]
	HELIX	37–47	0.314	0.326	0.826	0.760	0.950	
	RC	354–364	0.416	0.396	0.100	0.657	0.469	
1PLU		CHAIN	0.654	0.540	0.234	0.388	0.749	[10]
	HELIX	26–37	0.354	0.303	0.782	0.679	0.862	
	RC	302–313	0.617	0.536	0.284	0.489	0.668	

(Continued)

Table 11.1 Values of fuzzy oil drop parameters calculated for selected structures (structural units for which the 3D Gass function was defined) and "stop" fragments found within their sequences. Asterisks (*) indicate that the β-strand treated as stop fragment is an integral part of the solenoid First line describes the protein (identified by "chain")—for this unit the 3D Gasuss function was calculated; the second line (or more) describes status of polypeptide chain fragment treated as "stopper".—cont'd

Protein	Fragment		RD		Correlation coefficient			Ref.
	Structure	Residues	T-O-R	T-O-H	HvT	TvO	HvO	
2FKO		CHAIN	0.457	0.405	0.332	0.584	0.744	[11]
	BETA	1–6	0.207	0.257	0.487	0.896	0.633	
	BETA	137–144	0.400	0.163	0.531	0.652	0.794	
1QRM		CHAIN	0.475	0.471	0.363	0.661	0.737	[12]
	RC	5–9	0.591	0.214	0.164	0.961	0.311	
	HELIX	176–183	0.373	0.223	0.523	0.685	0.846	
1QRG		CHAIN	0.418	0.418	0.400	0.716	0.745	[12]
	RC	8–14	0.750	0.146	-0.317	0.712	0.138	
	HELIX	176–183	0.294	0.146	0.567	0.817	0.873	
1IDJ		CHAIN	0.722	0.650	0.216	0.313	0.737	[13]
	HELIX	27–36	0.255	0.258	0.727	0.868	0.937	
1OOC		CHAIN	0.643	0.501	0.238	0.429	0.704	[14]
	HELIX	39–50	0.369	0.296	0.809	0.716	0.829	
	RC	318–328	0.434	0.761	0.689	0.528	0.930	
1O88		CHAIN	0.650	0.541	0.230	0.392	0.748	[15]
	HELIX	25–37	0.382	0.325	0.781	0.644	0.856	
	RC	308–312	0.655	0.567	0.976	0.415	0.507	
1JRG		CHAIN	0.649	0.487	0.241	0.420	0.701	[16]
	HELIX	41–50	0.376	0.322	0.824	0.711	0.844	
	RC	320–327	0.457	0.848	0.574	0.459	0.970	
1JTA		CHAIN	0.632	0.489	0.240	0.440	0.706	[16]
	HELIX	40–51	0.402	0.344	0.832	0.650	0.811	
	RC	318–329	0.382	0.749	0.725	0.626	0.934	
2BSP		CHAIN	0.688	0.558	0.187	0.332	0.750	[17]
	HELIX	37–46	0.304	0.615	0.783	0.740	0.932	
	RC	349–364	0.450	0.463	0.241	0.637	0.625	

requirement, which is difficult to satisfy when designing β "stoppers". In addition, a putative short β-strand capable of arresting amyloid growth, may fail to retain its structural characteristics when isolated.

For the reasons stated above we believe that only a helical fragment, which is highly stable on its own, may serve as an efficient "stopper"—as long as it remains compatible with the target amyloid.

The relationship between "stop" fragments and the entirety of the molecule becomes even more transparent in the context of solenoids equipped

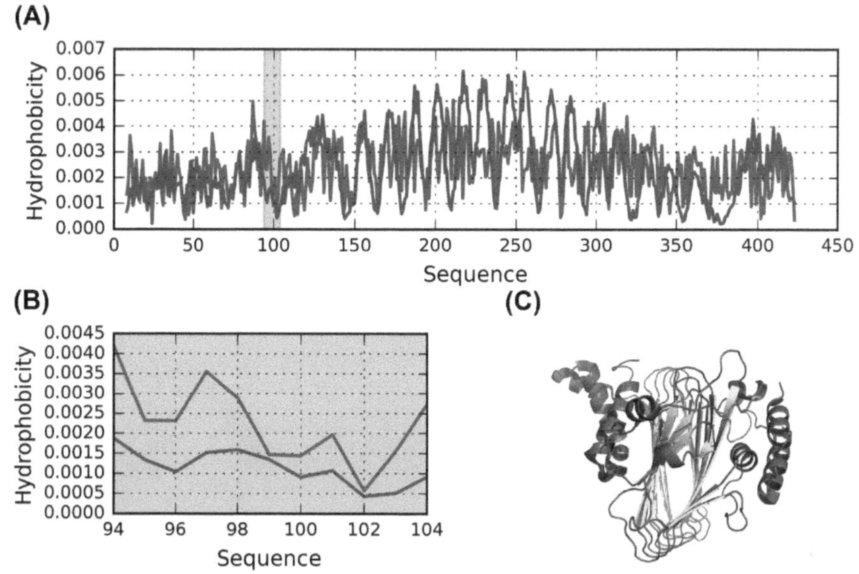

Fig. 11.1 FOD characteristic of sufc-sufd complex involved in the iron-sulfur cluster biosynthesis (2ZU0): (A)—hydrophobicity distribution profiles: T (blue), O (red); green background—"stopper(s)". (B)—detailed view of profiles from A, focused on "stop" fragment(s) only. (C)—3D presentation of the protein (green color—"stopper(s)", N-terminal in the foreground).

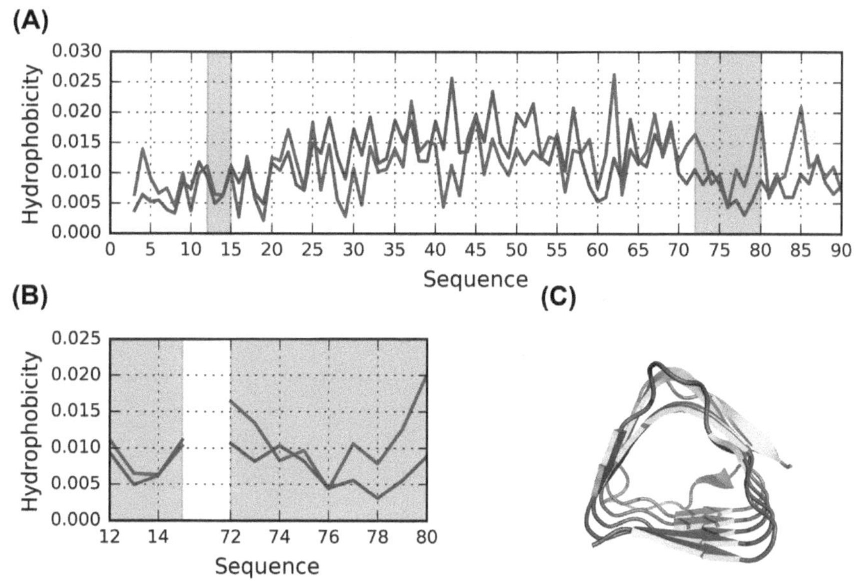

Fig. 11.2 FOD characteristic of antifreeze protein from *Choristoneura fumiferana* (1L0S): (A) hydrophobicity distribution profiles: T (blue), O (red); green background—" stopper(s)". (B) detailed view of profiles from A, focused on "stop" fragment(s) only. (C) 3D presentation of the protein (green color—"stopper(s)", N-terminal in the foreground).

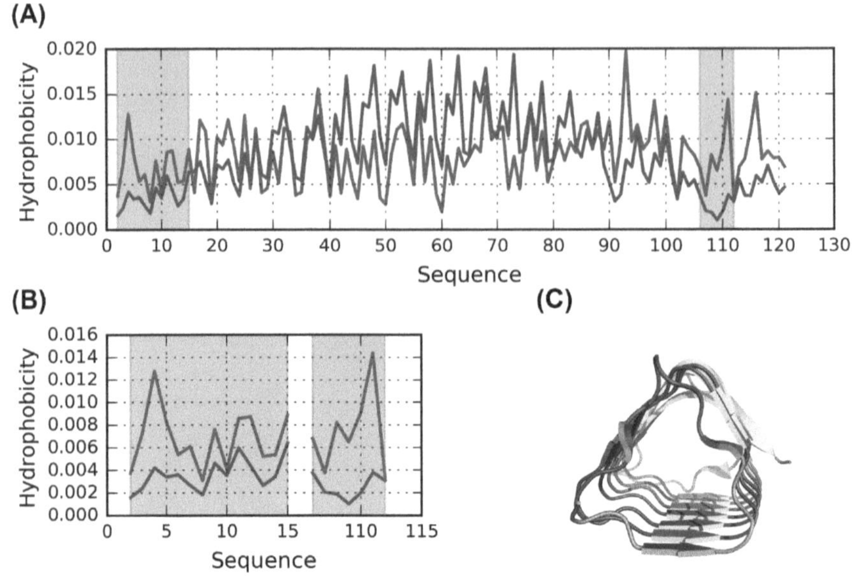

Fig. 11.3 FOD characteristic of antifreeze protein from *Choristoneura fumiferana* (1M8N): (A) hydrophobicity distribution profiles: T (blue), O (red); green background—" stopper(s)". (B) detailed view of profiles from A, focused on "stop" fragment(s) only. (C) 3D presentation of the protein (green color—"stopper(s)", N-terminal in the foreground).

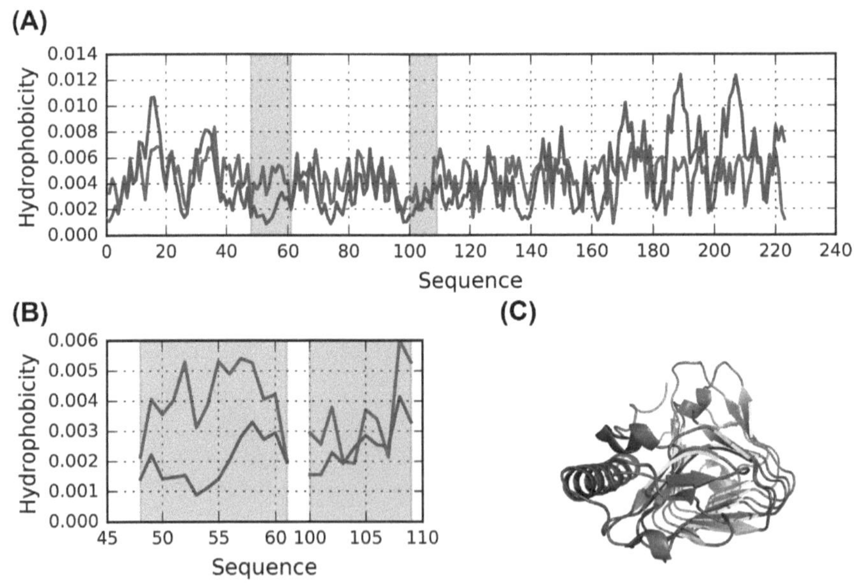

Fig. 11.4 FOD characteristic of fungal antifreeze protein from Typhula ishikariensis. (3VN3): (A) hydrophobicity distribution profiles: T (blue), O (red); green background—" stopper(s)". (B) detailed view of profiles from A, focused on "stop" fragment(s) only. (C) 3D presentation of the protein (green color—"stopper(s)", N-terminal in the foreground).

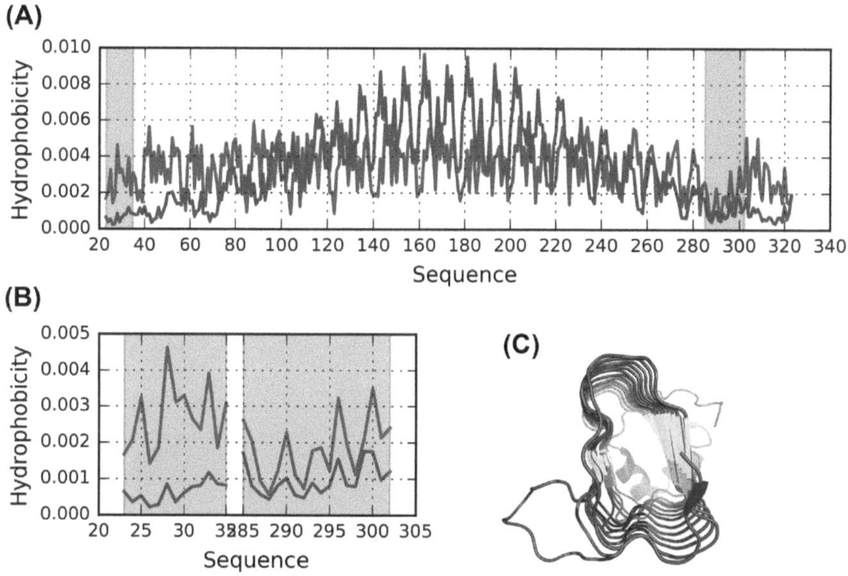

Fig. 11.5 FOD characteristic of bacterial antifreeze protein from *Marinomonas primoryensis* (3P4G): (A) hydrophobicity distribution profiles: T (blue), O (red); green background—" stopper(s)". (B) detailed view of profiles from A, focused on "stop" fragment(s) only. (C) 3D presentation of the protein (green color—"stopper(s)", N-terminal in the foreground).

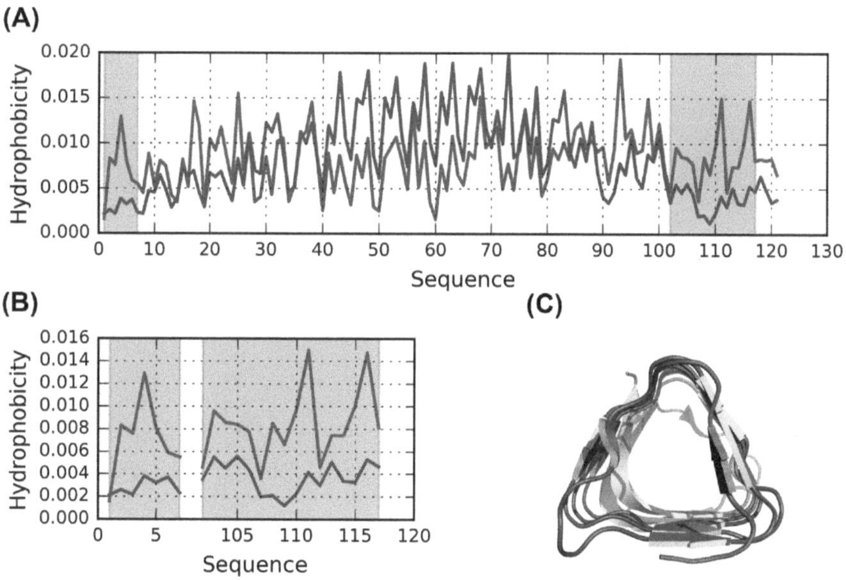

Fig. 11.6 FOD characteristic of antifreeze protein from spruce budworm (*Choristoneura fumiferana.*) (1Z2F): (A) hydrophobicity distribution profiles: T (blue), O (red); green background—"stopper(s)". (B) detailed view of profiles from A, focused on "stop" fragment(s) only. (C) 3D presentation of the protein (green color—"stopper(s)", N-terminal in the foreground).

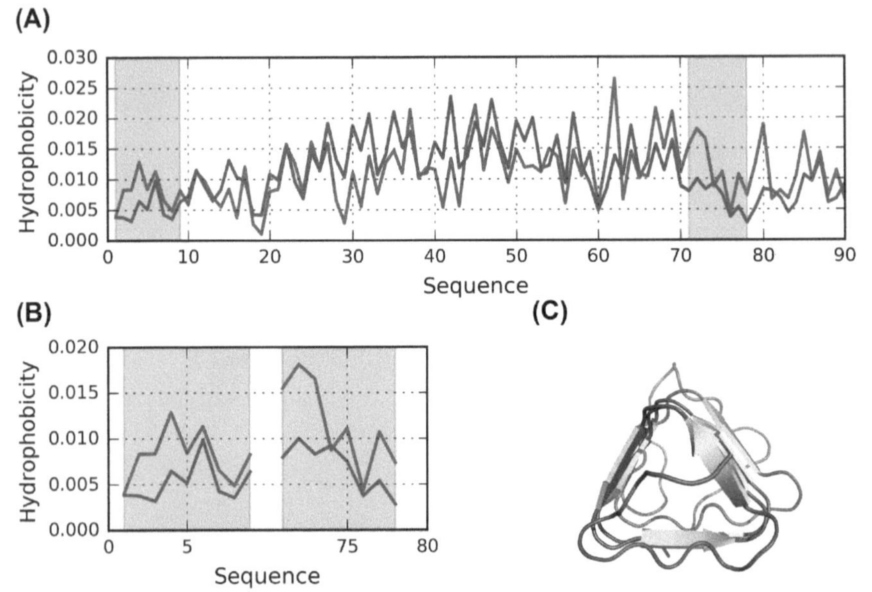

Fig. 11.7 FOD characteristic of antifreeze protein from Spruce budworm. (*Choristoneura fumiferana*). (1N4I): (A) hydrophobicity distribution profiles: T (blue), O (red); green background—"stopper(s)". (B) detailed view of profiles from A, focused on "stop" fragment(s) only. (C) 3D presentation of the protein (green color—"stopper(s)", N-terminal in the foreground).

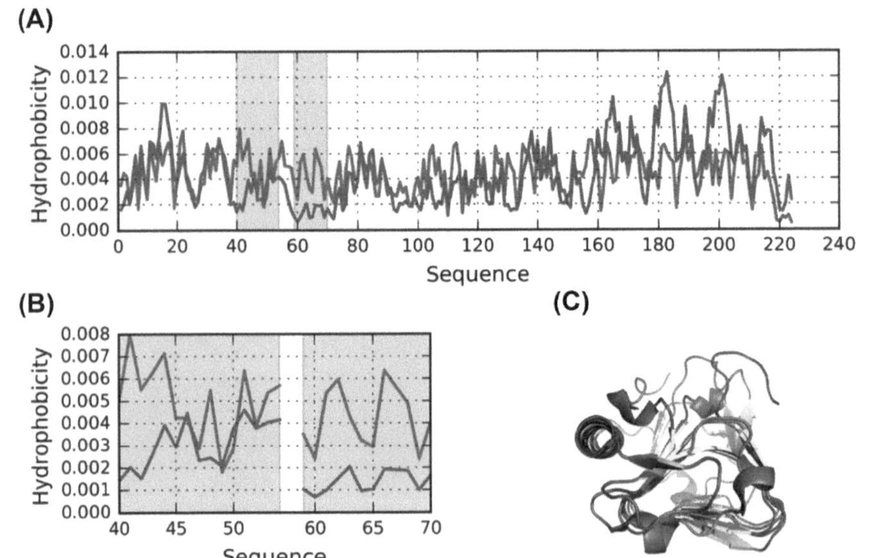

Fig. 11.8 FOD characteristic of antifreeze protein from an antarctic sea ice bacterium *colwellia sp* (3WP9): (A) hydrophobicity distribution profiles: T (blue), O (red); green background—"stopper(s)". (B) detailed view of profiles from A, focused on "stop" fragment(s) only. (C) 3D presentation of the protein (green color—"stopper(s)", N-terminal in the foreground).

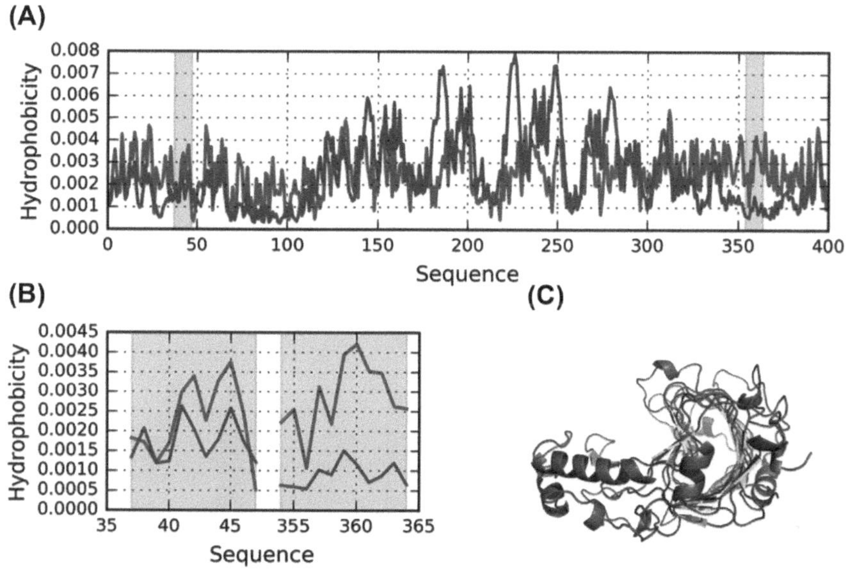

Fig. 11.9 FOD characteristic of Bacillus subtilis pectate lyase (1BN8): (A) hydrophobicity distribution profiles: T (blue), O (red); green background—"stopper(s)". (B) detailed view of profiles from A, focused on "stop" fragment(s) only. (C) 3D presentation of the protein (green color—"stopper(s)", N-terminal in the foreground).

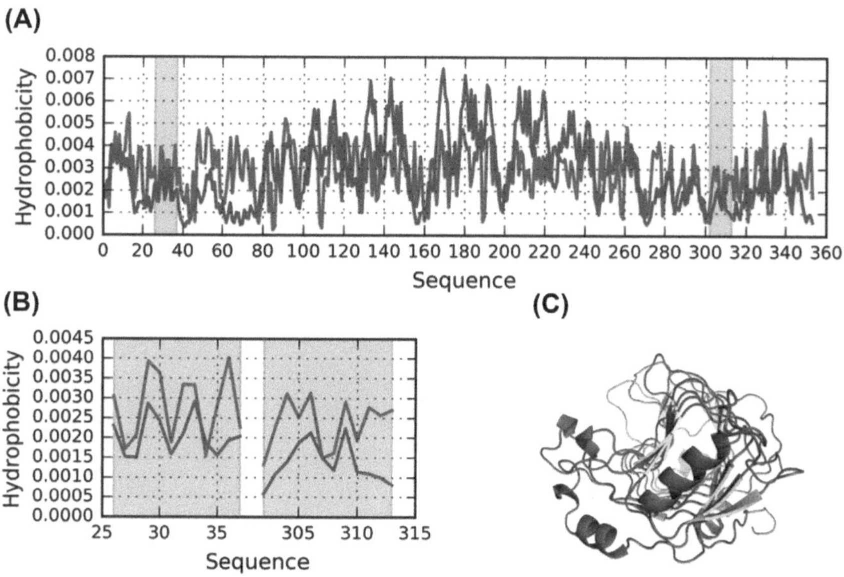

Fig. 11.10 FOD characteristic of pectate lyase C from erwinia chrysanthemi with 1 lu+3 ion in the putative calcium binding site (1PLU): (A) hydrophobicity distribution profiles: T (blue), O (red); green background—"stopper(s)". (B) detailed view of profiles from A, focused on "stop" fragment(s) only. (C) 3D presentation of the protein (green color—" stopper(s)", N-terminal in the foreground).

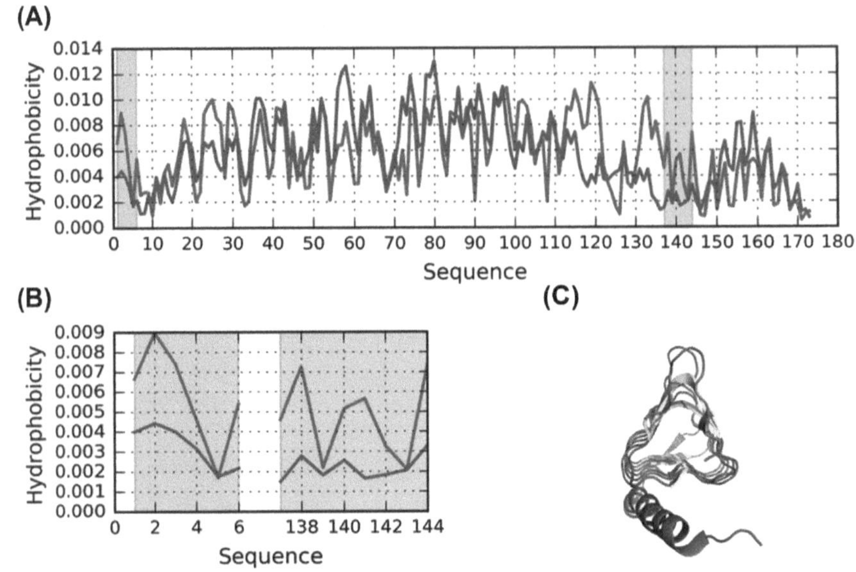

Fig. 11.11 FOD characteristic of carbonic anhydrase from *Pyrococcus horikoshii.* (2FKO): (A) hydrophobicity distribution profiles: T (blue), O (red); green background—" stopper(s)". (B) detailed view of profiles from A, focused on "stop" fragment(s) only. (C) 3D presentation of the protein (green color—"stopper(s)", N-terminal in the foreground).

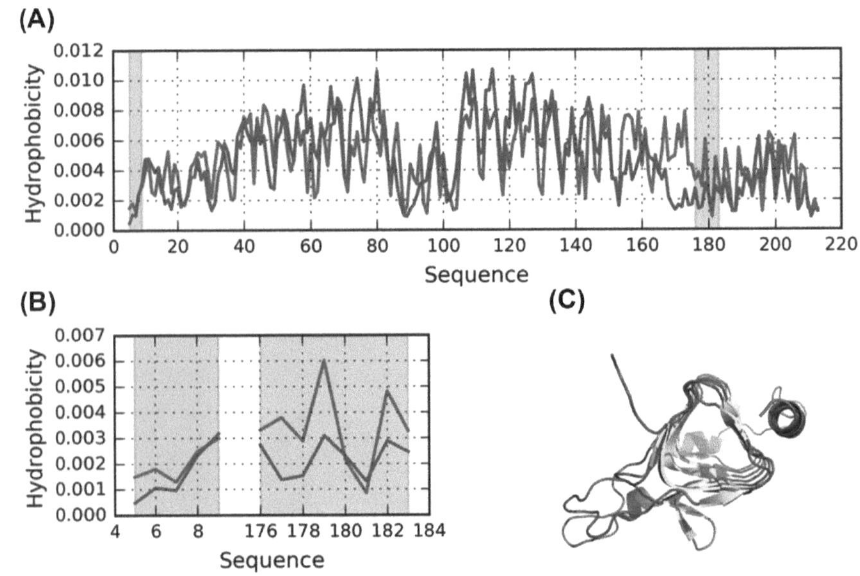

Fig. 11.12 FOD characteristic of carbonic anhydrase from *Methanosarcina thermophila* (1QRM): (A) hydrophobicity distribution profiles: T (blue), O (red); green background—" stopper(s)". (B) detailed view of profiles from A, focused on "stop" fragment(s) only. (C) 3D presentation of the protein (green color—"stopper(s)", N-terminal in the foreground).

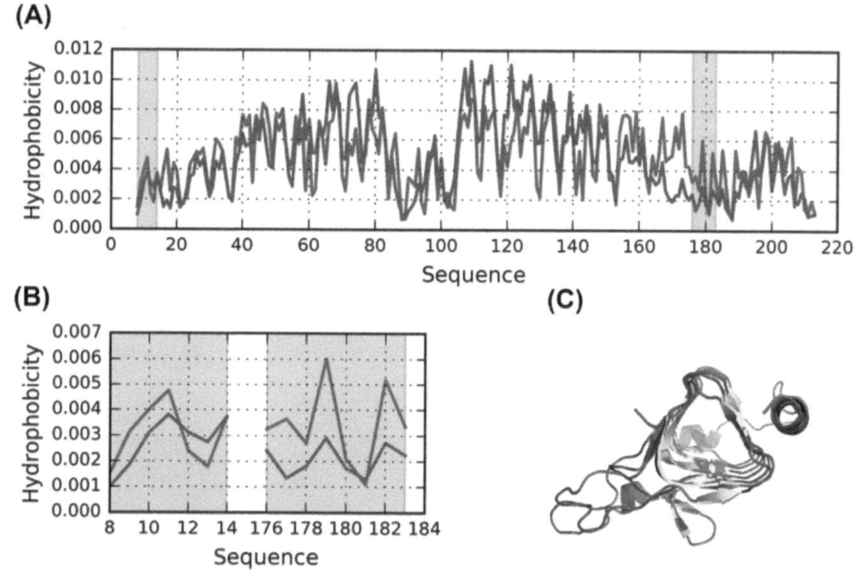

Fig. 11.13 FOD characteristic of carbonic anhydrase from *methanosarcina thermophila* (1QRG): (A) hydrophobicity distribution profiles: T (blue), O (red); green background—" stopper(s)". (B) detailed view of profiles from A, focused on "stop" fragment(s) only. (C) 3D presentation of the protein (green color—"stopper(s)", N-terminal in the foreground).

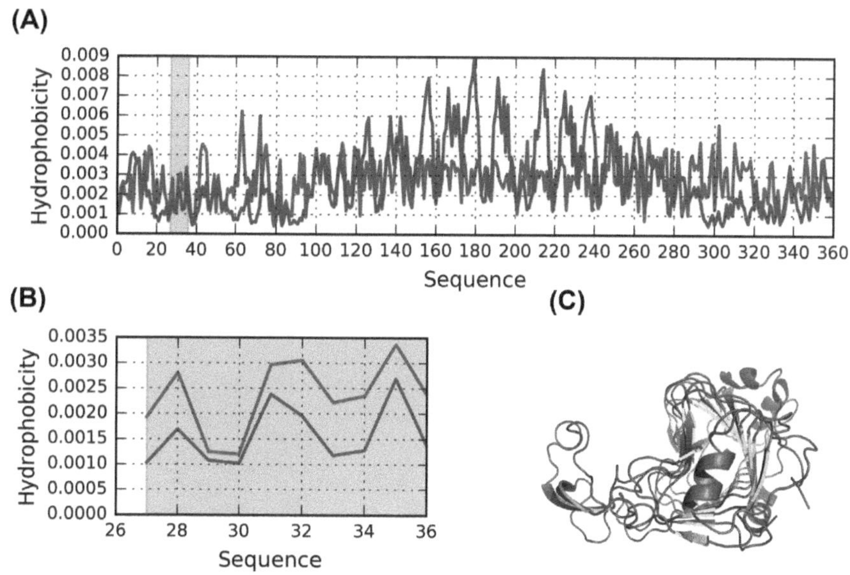

Fig. 11.14 FOD characteristic of pectin lyase from *Aspergillus niger*. (1IDJ): (A) hydrophobicity distribution profiles: T (blue), O (red); green background—" stopper(s)". (B) detailed view of profiles from A, focused on "stop" fragment(s) only. (C) 3D presentation of the protein (green color—"stopper(s)", N-terminal in the foreground).

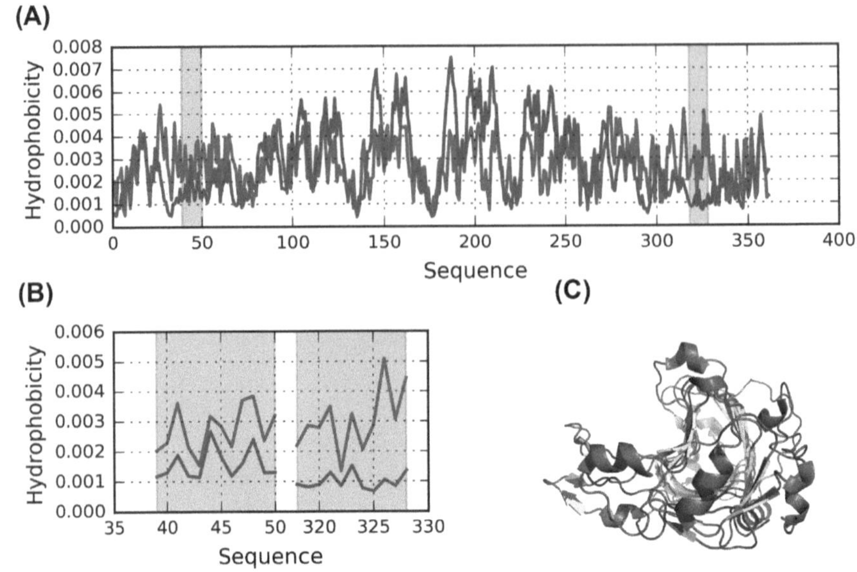

Fig. 11.15 FOD characteristic of pectate lyase from *Erwinia chrysanthem* (1OOC): (A) hydrophobicity distribution profiles: T (blue), O (red); green background—" stopper(s)". (B) detailed view of profiles from A, focused on "stop" fragment(s) only. (C) 3D presentation of the protein (green color—"stopper(s)", N-terminal in the foreground).

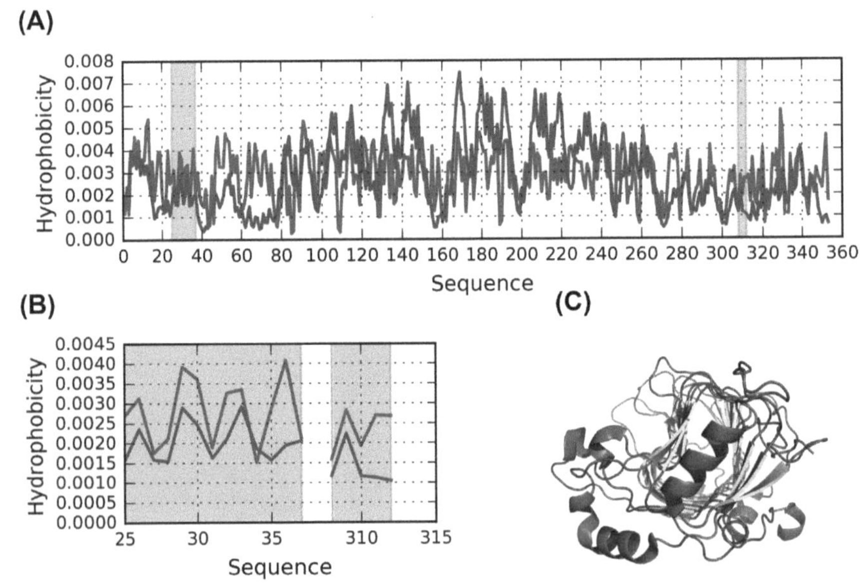

Fig. 11.16 FOD characteristic of pectate lyase C from *Erwinia chrysanthemi* (1O88): (A) hydrophobicity distribution profiles: T (blue), O (red); green background—" stopper(s)". (B) detailed view of profiles from A, focused on "stop" fragment(s) only. (C) 3D presentation of the protein (green color—"stopper(s)", N-terminal in the foreground).

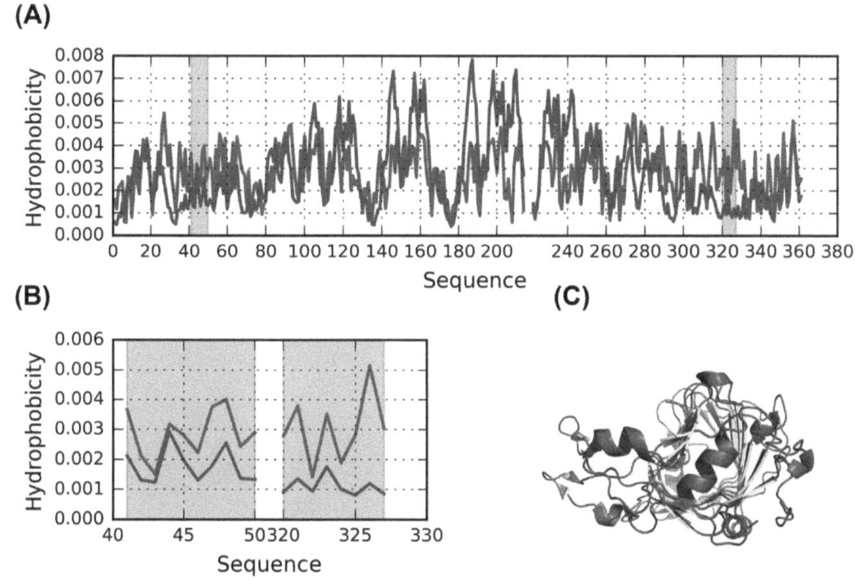

Fig. 11.17 FOD characteristic of pectate lyase from *Erwinia chrysanthemi* (1JRG): (A) hydrophobicity distribution profiles: T (blue), O (red); green background—" stopper(s)". (B) detailed view of profiles from A, focused on "stop" fragment(s) only. (C) 3D presentation of the protein (green color—"stopper(s)", N-terminal in the foreground).

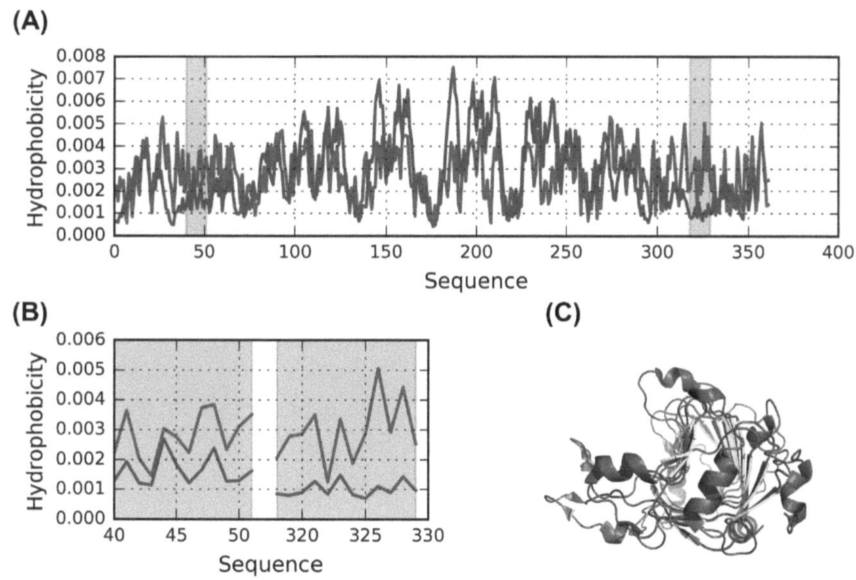

Fig. 11.18 FOD characteristic of pectate lyase from *Erwinia chrysanthemi*. (1JTA): (A) hydrophobicity distribution profiles: T (blue), O (red); green background—" stopper(s)". (B) detailed view of profiles from A, focused on "stop" fragment(s) only. (C) 3D presentation of the protein (green color—"stopper(s)", N-terminal in the foreground).

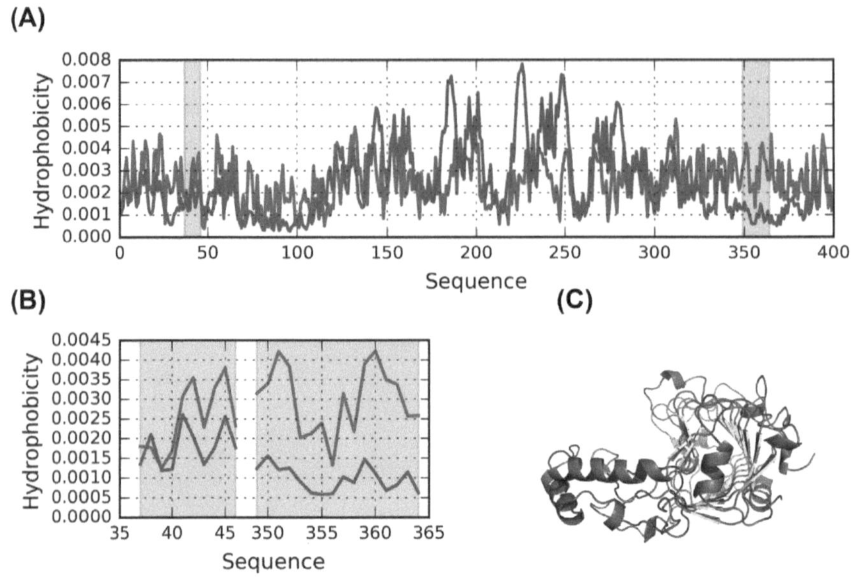

Fig. 11.19 FOD characteristic of pectate lyase from *Bacillus subtilis* (2BSP): (A) hydrophobicity distribution profiles: T (blue), O (red); green background—" stopper(s)". (B) detailed view of profiles from A, focused on "stop" fragment(s) only. (C) 3D presentation of the protein (green color—"stopper(s)", N-terminal in the foreground).

with such "stoppers". We discuss the structural properties of solenoid fragments in a separate chapter (see Chapter 7).

When discussing potential drugs capable of counteracting linear propagation of polypeptide chains (including amyloids), we should not neglect to acknowledge other proposals [20—40]. While the presented work focuses on peptide "stoppers", much research has been directed toward investigating organic compounds capable of meeting this goal [20—40]. Where peptides are mentioned, the authors usually focus their attention at β-strand—however, as noted above, designing such stoppers appears far more challenging than coming up with their helical equivalents. It is also worth noting—when looking at the contents of Table 11.1—that in all biological proteins the role of "caps" falls to helical fragments.

References

[1] Wada K, Sumi N, Nagai R, Iwasaki K, Sato T, Suzuki K, Fukuyama K. Molecular dynamism of Fe—S cluster biosynthesis implicated by the structure of the SufC2—SufD2 complex. Journal of Molecular Biology 2009;387(1):245—58. https://doi.org/10.1016/j.jmb.2009.01.054.

[2] Leinala EK, Davies PL, Jia Z. Crystal structure of β-helical antifreeze protein points to a general ice binding model. Structure 2002;10(5):619–27. https://doi.org/10.1016/s0969-2126(02)00745-1.
[3] Leinala EK. A beta -helical antifreeze protein isoform with increased activity. Structural and functional insights. Journal of Biological Chemistry 2002;277(36):33349–52. https://doi.org/10.1074/jbc.m205575200.
[4] Kondo H, Hanada Y, Sugimoto H, Hoshino T, Garnham CP, Davies PL, Tsuda S. Ice-binding site of snow mold fungus antifreeze protein deviates from structural regularity and high conservation. Proceedings of the National Academy of Sciences 2012; 109(24):9360–5. https://doi.org/10.1073/pnas.1121607109.
[5] Garnham CP, Campbell RL, Davies PL. Anchored clathrate waters bind antifreeze proteins to ice. Proceedings of the National Academy of Sciences 2011;108(18): 7363–7. https://doi.org/10.1073/pnas.1100429108.
[6] Li C, Guo X, Jia Z, Xia B, Jin C. Solution structure of an antifreeze protein CfAFP-501 from Choristoneura fumiferana. Journal of Biomolecular NMR 2005;32(3): 251–6. https://doi.org/10.1007/s10858-005-8206-3.
[7] Graether SP, Gagné SM, Spyracopoulos L, Jia Z, Davies PL, Sykes BD. Spruce budworm antifreeze protein: changes in structure and dynamics at low temperature. Journal of Molecular Biology 2003;327(5):1155–68. https://doi.org/10.1016/s0022-2836(03)00235-3.
[8] Hanada Y, Nishimiya Y, Miura A, Tsuda S, Kondo H. Hyperactive antifreeze protein from an Antarctic sea ice bacterium Colwelliasp has a compound ice-binding site without repetitive sequences. FEBS Journal 2014;281(16):3576–90. https://doi.org/10.1111/febs.12878.
[9] Pickersgill R, Jenkins J, Harris G, Nasser W, Robert-Baudouy J. The structure of Bacillus subtilis pectate lyase in complex with calcium. Nature Structural & Molecular Biology 1994;1(10):717–23. https://doi.org/10.1038/nsb1094-717.
[10] Yoder MD, Jurnak F. The refined three-dimensional structure of pectate lyase C from erwinia chrysanthemi at 2.2 Angstrom resolution (implications for an enzymatic mechanism). Plant Physiology 1995;107(2):349–64. https://doi.org/10.1104/pp.107.2.349.
[11] Jeyakanthan J, Rangarajan S, Mridula P, Kanaujia SP, Shiro Y, Kuramitsu S, Sekar K. Observation of a calcium-binding site in the γ-class carbonic anhydrase from Pyrococcus horikoshii. Acta Crystallographica Section D Biological Crystallography 2008; 64(10):1012–9. https://doi.org/10.1107/s0907444908024323.
[12] Iverson TM, Alber BE, Kisker C, Ferry JG, Rees DC. A closer look at the active site of γ-class carbonic anhydrases: high-resolution crystallographic studies of the carbonic anhydrase from Methanosarcina thermophila. Biochemistry 2000;39(31):9222–31. https://doi.org/10.1021/bi000204s.
[13] Mayans O, Scott M, Connerton I, Gravesen T, Benen J, Visser J, Jenkins J. Two crystal structures of pectin lyase A from Aspergillus reveal a pH driven conformational change and striking divergence in the substrate-binding clefts of pectin and pectate lyases. Structure 1997;5(5):677–89. https://doi.org/10.1016/s0969-2126(97)00222-0.
[14] Dehdashti SJ, Doan CN, Chao KL, Yoder MD. Effect of mutations in the T1.5 loop of pectate lyase A from Erwinia chrysanthemi EC16. Acta Crystallographica Section D Biological Crystallography 2003;59(7):1339–42. https://doi.org/10.1107/s0907444903011491.
[15] Herron SR, Scavetta RD, Garrett M, Legner M, Jurnak F. Characterization and implications of Ca^{2+} Binding to pectate lyase C. Journal of Biological Chemistry 2003; 278(14):12271–7. https://doi.org/10.1074/jbc.m209306200.

[16] Thomas LM, Doan CN, Oliver RL, Yoder MD. Structure of pectate lyase A: comparison to other isoforms. Acta Crystallographica Section D Biological Crystallography 2002;58(6):1008–15. https://doi.org/10.1107/s0907444902005851.

[17] Pickersgill, R. Bacillus subtilis pectate lyase r279k mutant. https://www.rcsb.org/structure/2BSP.

[18] Roterman I, Banach M, Konieczny L. Propagation of fibrillar structural forms in proteins stopped by naturally occurring short polypeptide chain fragments. Pharmaceuticals 2017;10(4):E89. https://doi.org/10.3390/ph10040089. pii.

[19] Roterman I, Banach M, Konieczny L. Towards the design of anti-amyloid short peptide helices. Bioinformation 2018;14(1):1–7. https://doi.org/10.6026/97320630014001. eCollection 2018.

[20] Baggett DW, Nath A. The rational discovery of a tau aggregation inhibitor. Biochemistry 2018;57(42):6099–107. https://doi.org/10.1021/acs.biochem.8b00581.

[21] Becker M, Moore A, Naughton M, Boland B, Siems W-E, Walther T. Neprilysin degrades murine Amyloid-β (Aβ) more efficiently than human Aβ: further implication for species-specific amyloid accumulation. Neuroscience Letters 2018;686:74–9. https://doi.org/10.1016/j.neulet.2018.08.028.

[22] Brumshtein B, Esswein SR, Salwinski L, Phillips ML, Ly AT, Cascio D, Eisenberg DS. Inhibition by small-molecule ligands of formation of amyloid fibrils of an immunoglobulin light chain variable domain. eLife 2015;4. https://doi.org/10.7554/elife.10935.

[23] Cao Q, Shin WS, Chan H, Vuong CK, Dubois B, Li B, Jiang L. Inhibiting amyloid-β cytotoxicity through its interaction with the cell surface receptor LilrB2 by structure-based design. Nature Chemistry 2018;10(12):1213–21. https://doi.org/10.1038/s41557-018-0147-z.

[24] Cox D, Whiten DR, Brown JWP, Horrocks MH, San Gil R, Dobson CM, Ecroyd H. The small heat shock protein Hsp27 binds α-synuclein fibrils, preventing elongation and cytotoxicity. Journal of Biological Chemistry 2018;293(12):4486–97. https://doi.org/10.1074/jbc.m117.813865.

[25] Cremades N, Dobson CM. The contribution of biophysical and structural studies of protein self-assembly to the design of therapeutic strategies for amyloid diseases. Neurobiology of Disease 2018;109:178–90. https://doi.org/10.1016/j.nbd.2017.07.009.

[26] Davis IW, Raha K, Head MS, Baker D. Blind docking of pharmaceutically relevant compounds using RosettaLigand. Protein Science 2009;18(9):1998–2002. https://doi.org/10.1002/pro.192.

[27] Filippi L, Chiaravalloti A, Bagni O, Schillaci O. ^{18}F-labeled radiopharmaceuticals for the molecular neuroimaging of amyloid plaques in Alzheimer's disease. American Journal of Nuclear Medicine and Molecular Imaging 2018;8(4):268–81.

[28] Giorgetti S, Greco C, Tortora P, Aprile F. Targeting amyloid aggregation: an overview of strategies and mechanisms. International Journal of Molecular Sciences 2018;19(9):2677. https://doi.org/10.3390/ijms19092677.

[29] Gorantla NV, Chinnathambi S. Tau protein squired by molecular chaperones during alzheimer's disease. Journal of Molecular Neuroscience 2018;66(3):356–68. https://doi.org/10.1007/s12031-018-1174-3.

[30] Kundel F, De S, Flagmeier P, Horrocks MH, Kjaergaard M, Shammas SL, Klenerman D. Hsp70 inhibits the nucleation and elongation of tau and sequesters tau aggregates with high affinity. ACS Chemical Biology 2018;13(3):636–46. https://doi.org/10.1021/acschembio.7b01039.

[31] Kurnik M, Sahin C, Andersen CB, Lorenzen N, Giehm L, Mohammad-Beigi H, Otzen DE. Potent α-synuclein aggregation inhibitors, identified by high-

throughput screening, mainly target the monomeric state. Cell Chemical Biology 2018;25(11):1389–402. https://doi.org/10.1016/j.chembiol.2018.08.005. e9.
[32] Markowicz-Piasecka M, Huttunen KM, Sikora J. Metformin and its sulphonamide derivative simultaneously potentiateanti-cholinesterase activity of donepezil and inhibit beta-amyloid aggregation. Journal of Enzyme Inhibition and Medicinal Chemistry 2018;33(1):1309–22. https://doi.org/10.1080/14756366.2018.1499627.
[33] Podder A, Pandit M, Narayanan L. Drug target prioritization for alzheimer's disease using protein interaction network analysis. OMICS 2018;22(10):665–77. https://doi.org/10.1089/omi.2018.0131.
[34] Pujols J, Peña-Díaz S, Lázaro DF, Peccati F, Pinheiro F, González D, Ventura S. Small molecule inhibits α-synuclein aggregation, disrupts amyloid fibrils, and prevents degeneration of dopaminergic neurons. Proceedings of the National Academy of Sciences 2018;115(41):10481–6. https://doi.org/10.1073/pnas.1804198115.
[35] Ramesh M, Makam P, Voshavar C, Khare H, Rajasekhar K, Ramakumar S, Govindaraju T. l-Dopa and dopamine conjugated naphthalenediimides modulate amyloid β toxicity. Organic and Biomolecular Chemistry 2018;16(41):7682–92. https://doi.org/10.1039/c8ob01691g.
[36] Sanjeev A, Mattaparthi VSK. Computational study on the role of γ-synuclein in inhibiting the α-synuclein aggregation. Central Nervous System Agents in Medicinal Chemistry 2018;18. https://doi.org/10.2174/1871524918666181012160439.
[37] Sarnataro D. Attempt to untangle the prion-like misfolding mechanism for neurodegenerative diseases. International Journal of Molecular Sciences 2018;19(10):3081. https://doi.org/10.3390/ijms19103081.
[38] Seidler PM, Boyer DR, Rodriguez JA, Sawaya MR, Cascio D, Murray K, Eisenberg DS. Structure-based inhibitors of tau aggregation. Nature Chemistry 2017;10(2):170–6. https://doi.org/10.1038/nchem.2889.
[39] Skolnick J, Gao M, Roy A, Srinivasan B, Zhou H. Implications of the small number of distinct ligand binding pockets in proteins for drug discovery, evolution and biochemical function. Bioorganic & Medicinal Chemistry Letters 2015;25(6):1163–70. https://doi.org/10.1016/j.bmcl.2015.01.059.
[40] Zhou H, Gao M, Skolnick J. Comprehensive prediction of drug-protein interactions and side effects for the human proteome. Scientific Reports 2015;5(1). https://doi.org/10.1038/srep11090.

CHAPTER 12

The hypothetical amyloid transformation of transthyretin

Mateusz Banach[1], **Leszek Konieczny**[2], **Irena Roterman**[1]
[1]Department of Bioinformatics and Telemedicine, Jagiellonian University – Medical College, Krakow, Poland
[2]Chair of Medical Biochemistry, Jagiellonian University – Medical College, Krakow, Poland

Contents

Conceptual diagram presenting linear propagation of (gray) bands with varying hydrophobicity. If this linear pattern is joined by fragments marked in red (initially not fitting to linear propagation), the protein is at risk of undergoing amyloid transformation.

If we accept the presented amyloid identification criteria as accurate, we may apply them to predict the structures of various amyloids – including the transthyretin amyloid. The structure of this protein is listed in PDB under ID 1DVQ [1], in its dimeric form, with each chain consisting of residues 10–124. Its structure is characterized as a mainly β-sandwich (CATH code 2.60.40.180). Each β-sheet consists of four antiparallel β-strands. In addition, a single helical fragment is also present (residues 74–81).

Fig. 12.1 illustrates the distributions of hydrophobicity (T, O and H) as it appears in both chains treated as parts of dimer and as individual structural units. The aim of Fig. 12.1 is to visualize the relation between T and O as well as between O and H, with some fragments exhibiting strong alignment between the observed and intrinsic distributions. The two chains appear almost identical in terms of all three distributions.

From Globular Proteins to Amyloids
ISBN: 978-0-08-102981-7
https://doi.org/10.1016/B978-0-08-102981-7.00020-8

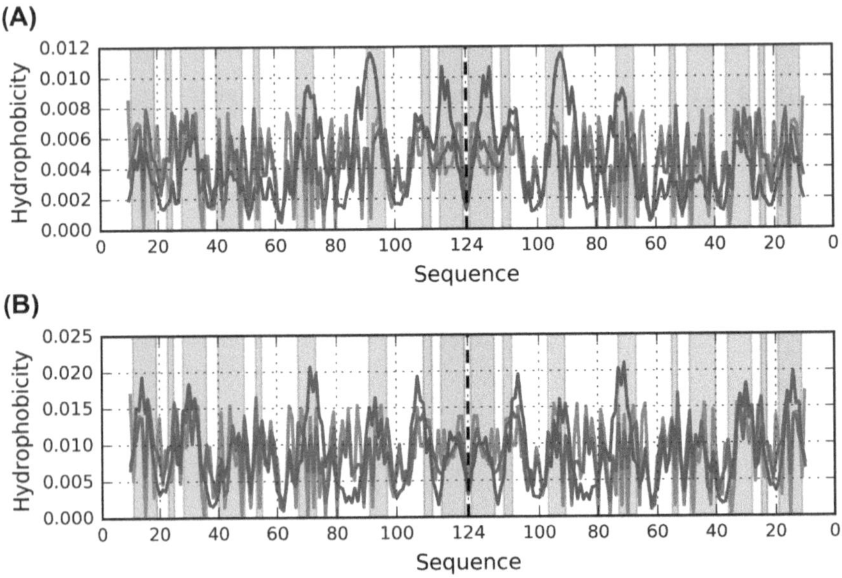

Fig. 12.1 Theoretical (T, blue) and observed (O, red) hydrophobicity distribution profiles for transthyretin dimer (1DVQ) chains A and B (A) treated as part of the dimer. (B) treated as individual units.

Thick dashed line at position 124 separates chain A (left) from chain B (right). To reproduce the symmetry of the complex, results for chain B are present in reverse order. Components of the appropriate β-sheets are marked by red (starting strand 28–36, called also as II) and blue (starting strand 11–19, called also as I) backgrounds.

Clearly, the structure lacks a prominent hydrophobic core (Fig. 12.1 and Table 12.1). The RD (T-O-R) higher than 0.5 expresses the lack of centric hydrophobic core. The RD (T-O-H) below 0.5 suggests not significant influence of intrinsic hydrophobicity. However the highest correlation coefficient for HvO relation suggest the strong influence of intrinsic hydrophobicity on the final distribution. The lack of centric hydrophobic core seems to be caused mainly by the fragments at 60–80, 80–100 and 100–120 (Fig. 12.1). Additionally, as suggested by the arrangement of β-strands, it appears that the initial β-strand at 11–19 (marked on Fig. 12.1 by blue background) is closer to the theoretical distribution whereas the strand at 28–36 (red background on Fig. 12.1) is more discordant. This observation is supported by the parameters listed in Table 12.1, where the β-sheet 28–36 (the starting fragment used to identify this sheet)

Table 12.1 Fuzzy oil drop parameters (RD values and correlation coefficients) for transthyretin (1DVQ): dimer, chains treated as parts of dimer, chains treated as individual units and β-sheets comprising the dimer (with status of β-strands as they appear in those sheets). Values given in bold identify discordance (RD > 0.5) and positions with large disproportion of correlation coefficient.

Trans thyretin (1DVQ)	Fragment	RD		Correlation coefficient		
		T-O-R	T-O-H	HvT	TvO	HvO
Dimer	A + B	**0.650**	0.467	**0.222**	**0.368**	**0.722**
	Chain A in dimer	**0.652**	0.467	**0.225**	**0.352**	**0.723**
	Chain B in dimer	**0.648**	0.468	**0.220**	**0.380**	**0.720**
	Chain A individual	**0.562**	0.370	0.328	0.592	0.687
	Chain B individual	**0.556**	0.368	0.317	0.591	0.687
β-strands						
	11–19	0.440	0.089	0.589	0.557	0.853
	23–25 + 53–55	0.411	0.343	0.682	0.602	0.751
	104–112	0.148	0.184	0.677	0.910	0.586
	115–123	**0.639**	**0.697**	0.054	0.576	0.568
β-sheet I		**0.717**	0.444	0.251	0.274	0.582
β-strands						
	28–36	0.467	0.171	0.468	0.489	0.872
	40–49	**0.517**	0.216	0.713	0.155	0.533
	67–73	**0.760**	0.182	**−0.263**	**−0.549**	**0.898**
	91–97	**0.776**	**0.704**	**−0.090**	**−0.423**	**0.897**
β-sheet II		**0.682**	0.326	**0.162**	**0.144**	**0.764**

appears to be strongly dominated by the intrinsic hydrophobicity of its component residues.

In search for fragments representing status recognized as amyloidogenic (Chapter 10), the profile of correlation coefficients (for 5 aa window) is presented in Fig. 12.2. Based on the description and applicability of the fuzzy oil drop model discussed in previous chapters, we may draw conclusions regarding the propensity of the presented structures to undergo amyloid transformation. This process is aided by identifying sections where the observed distribution deviates from the theoretical distribution in specific ways. We therefore calculate correlation coefficients for consecutive 5 aa fragments (using an overlapping moving frame), seeking fragments which exhibit low (preferably negative) values of HvT and TvO, coupled with

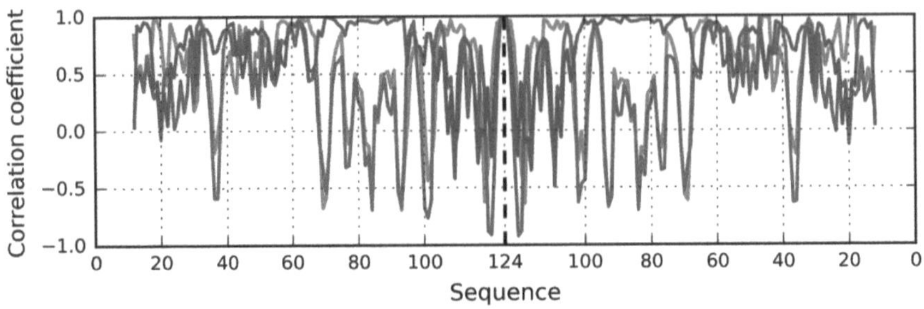

Fig. 12.2 Correlation coefficients between FOD hydrophobicity profiles calculated for transthyretin dimer (1DVQ) in 5 aa overlapping moving frame mode: HvO – blue, HvT – red, TvO – green. Thick dashed line at position 124 separates chain A (left) from chain B (right). To reproduce the symmetry of the complex, results for chain B are present in reverse order. This diagram enables us to identify amyloid transformation seeds (positions with disproportionally high HvO values vs. low HvT and TvO values).

high values of HvO. Fig. 12.2 shows the variability of correlation coefficients computed for each β-sheet in transthyretin.

Based on the chart shown in Fig. 12.2, the following fragments are seen to fulfill all amyloid identification criteria: 34–45, 66–69, 73–75, 80–84, 89–93 and 98–102 (in each chain). To better visualize the structural specificity, the correlation coefficients limited to β-sheets following the spatial localization of certain polypeptide chain fragments is shown in Fig. 12.3.

Ordering of residues is consistent with their respective placement in the dimer (antiparallel arrangement). To avoid displaying multiple gaps between the strands, results are conflated into contiguous sequences, hence no exact numbering is marked on the horizontal axis.

Blue – HvO, red-TvO, green HvT.

Focusing the analysis on the status of β-sheets expressed by hydrophobicity profile correlation coefficients provides an unobstructed view of differences between those fragments. In the presented diagram (Fig. 12.3), amino acid sequences are ordered in accordance with the supersecondary structure of the dimer (this particularly concerns the ordering of β-strands in β-sheets). For this structure we calculate correlation coefficients using a 5 aa moving frame for β-sheets (treating them as the independent structural unit). The results (Fig. 12.3) reveal significant variability of β-sheets which make up the β-sandwich.

The sheet identified as "11–19" (I) does not appear to satisfy our amyloid identification criteria (Fig. 12.3B), whereas the sheet identified as

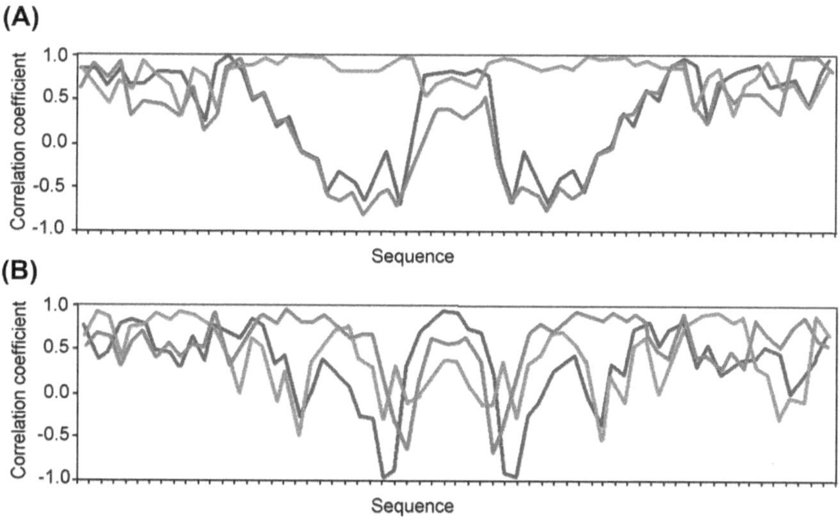

Fig. 12.3 Correlation coefficients between FOD hydrophobicity profiles calculated for β-sheets in transthyretin dimer (1DVQ) in 5-aa overlapping moving frame mode. (A) profile for β-sheet I (starting fragment 11–19). (B) profile for β-sheet II (starting fragment 28–36).

"28–36" (II) reveals strong involvement of fragments in which the observed distribution deviates from T in favor of H (Fig. 12.3A). Terminal residues (i.e. the initial and final fragments shown in Fig. 12.3A) are generally accordant with T. These residues correspond to the fragment at 40–49, which occupies an outlying location in the β-sheet. The fragments at 67–73 and 91–97 exhibit nonuniform correlation coefficients, (central section in Fig. 12.3A, showing negative values of HvT and TvO).

Assembling all the previously mentioned fragments (colored red in Fig. 12.4) identifies part of the β-sheet which – according to FOD based parameters – represents amyloid-like status. The loops (blue on Fig. 12.4) do not exhibit linear propagation of bands of different hydrophobicity. However if these two loops change their orientation versus the amyloid-like fragment of β-sheet, and continue the profile making it similar to the red fragment, the possible propagation in both directions appears to be possible.

It can be surmised that the fragments identified in the analysis of charts plotted in Fig. 12.3B act as amyloid seeds, since their spatial arrangement is already characteristic of a linear pattern. When joined by additional fragments (colored red) which do not belong to the β-sandwich, this pattern

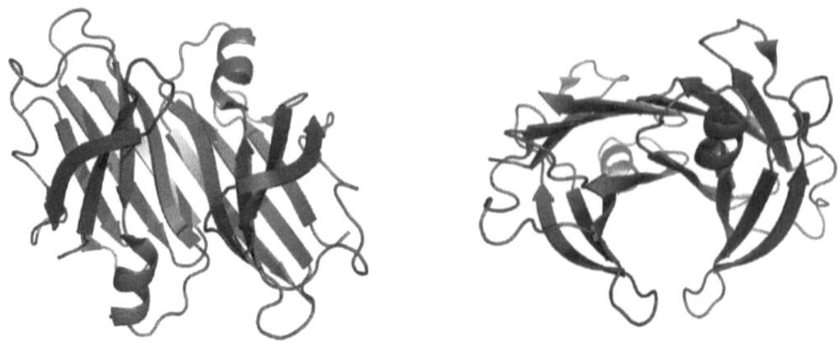

Fig. 12.4 3D presentation of transthyretin dimer (1DVQ) from two angles. Red fragments (67–73, 91–97, in each chain) express negative values of HvT and TvO, along with strongly positive values of HvO. Fragments 27–50 in each chain (shown in blue) are expected to undergo structural rearrangement in the process of amyloid transformation, aligning themselves with the existing linear pattern exhibited by other strands of β-sheet II.

spreads to the entire dimer. It also creates an open-ended structure which may attach itself to similarly arranged fragments in another dimer (and so on), ultimately producing a fibrillar structure. The structure illustrated in Fig. 12.5 also clearly shows a twist, which is characteristic of a fibril. Additional units (in this case — dimers) are complexed at an angle, resulting in a twisted structure.

Based on the presented correlation coefficients we may speculate that in the amyloid form of transthryretin residues 40–49 along with the entire fragment at 27–50 should adopt a linear structural pattern with other strands comprising β-sheet II. Such arrangement (which, of course, calls for additional conformational changes) may provide the necessary conditions for further propagation of the amyloid.

A similar structure, based on structural analysis of immunoglobulin light chain domain VL (relatively easy to undergo the amyloid transformation), has been proposed in Ref. [2]. While it differs somewhat to the one shown above, it also demonstrates the necessary conditions for further linear propagation in the event of certain rearrangements within the dimer. In this specific case, a bend in the fragment at 27–50 would enable it to align itself with the planar β-structure, facilitating further propagation of the observed discordance between T and O and preventing the dimer from forming a hydrophobic core.

Number of papers discussing transthyretin amyloid transformation is very large [3–17]. Among them are many of experimental character. Some of them discuss the possible therapeutic treatments [18–22].

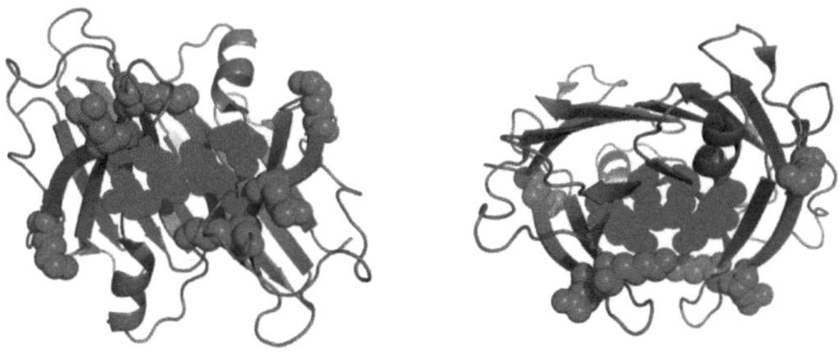

Fig. 12.5 3D presentation of transthyretin dimer (1DVQ) from two angles with residues forming a linear propagation of hydrophobicity distinguished in red (70, 72, 92) and residues exhibiting similar hydrophobic characteristics (low hydrophobicity with neighbors of higher hydrophobicity) distinguished in blue (35, 42 and 48). If the strands in 27–50 residue range (colored blue in Fig. 12.4) change the orientation to participate in linear propagation of low hydrophobicity band together with residues given in Fig. 12.4 in red, the unlimited propagation of bands of low and higher hydrophobicity bands on both sites (and thus amyloid formation) becomes possible.

References

[1] Klabunde T, Petrassi HM, Oza VB, Raman P, Kelly JW, Sacchettini JC. Rational design of potent human transthyretin amyloid disease inhibitors. Nature Structural Biology 2000;7(4):312–21. https://doi.org/10.1038/74082.

[2] Banach M, Kalinowska B, Konieczny L, Roterman I. Possible mechanism of amyloidogenesis of V domains. Self-Assembled Molecules – New Kind of Protein Ligands 2017:77–100. https://doi.org/10.1007/978-3-319-65639-7_5.

[3] Colon W, Kelly JW. Partial denaturation of transthyretin is sufficient for amyloid fibril formation in vitro. Biochemistry 1992;31(36):8654–60. https://doi.org/10.1021/bi00151a036.

[4] Connelly S, Choi S, Johnson SM, Kelly JW, Wilson IA. Structure-based design of kinetic stabilizers that ameliorate the transthyretin amyloidoses. Current Opinion in Structural Biology 2010;20(1):54–62. https://doi.org/10.1016/j.sbi.2009.12.009.

[5] Foss TR, Wiseman RL, Kelly JW. The pathway by which the tetrameric protein transthyretin dissociates. Biochemistry 2005;44(47):15525–33. https://doi.org/10.1021/bi051608t.

[6] Hurshman AR, White JT, Powers ET, Kelly JW. Transthyretin aggregation under partially denaturing conditions is a downhill polymerization. Biochemistry 2004;43(23):7365–81. https://doi.org/10.1021/bi049621l.

[7] Kelly JW. The alternative conformations of amyloidogenic proteins and their multi-step assembly pathways. Current Opinion in Structural Biology 1998;8(1):101–6. https://doi.org/10.1016/s0959-440x(98)80016-x.

[8] Kelly JW. Amyloid fibril formation and protein misassembly: a structural quest for insights into amyloid and prion diseases. Structure 1997;5(5):595–600. https://doi.org/10.1016/s0969-2126(97)00215-3.

[9] Kelly J, Colon W, Lai Z, Lashuel H, Mcculloch J, Mccutchen S, Peterson S. Transthyretin quaternary and tertiary structural changes facilitate misassembly into amyloid. Advances in Protein Chemistry 1997;50:161–81. https://doi.org/10.1016/s0065-3233(08)60321-6.

[10] Kelly JW. Alternative conformations of amyloidogenic proteins govern their behavior. Current Opinion in Structural Biology 1996;6(1):11–7. https://doi.org/10.1016/s0959-440x(96)80089-3.

[11] Knowles TPJ, Vendruscolo M, Dobson CM. The amyloid state and its association with protein misfolding diseases. Nature Reviews Molecular Cell Biology 2014;15(6):384–96. https://doi.org/10.1038/nrm3810.

[12] Lim KH, Dasari AKR, Hung I, Gan Z, Kelly JW, Wemmer DE. Structural changes associated with transthyretin misfolding and amyloid formation revealed by solution and solid-state NMR. Biochemistry 2016;55(13):1941–4. https://doi.org/10.1021/acs.biochem.6b00164.

[13] Lim KH, Dasari AKR, Ma R, Hung I, Gan Z, Kelly JW, Fitzgerald MC. Pathogenic mutations induce partial structural changes in the native β-sheet structure of transthyretin and accelerate aggregation. Biochemistry 2017;56(36):4808–18. https://doi.org/10.1021/acs.biochem.7b00658.

[14] Palaninathan SK, Mohamedmohaideen NN, Snee WC, Kelly JW, Sacchettini JC. Structural insight into pH-induced conformational changes within the native human transthyretin tetramer. Journal of Molecular Biology 2008;382(5):1157–67. https://doi.org/10.1016/j.jmb.2008.07.029.

[15] Saelices L, Chung K, Lee JH, Cohn W, Whitelegge JP, Benson MD, Eisenberg DS. Amyloid seeding of transthyretin by ex vivo cardiac fibrils and its inhibition. Proceedings of the National Academy of Sciences of the United States of America 2018;115(29):E6741–50. https://doi.org/10.1073/pnas.1805131115.

[16] Saelices L, Sievers SA, Sawaya MR, Eisenberg DS. Crystal structures of amyloidogenic segments of human transthyretin. Protein Science 2018;27(7):1295–303. https://doi.org/10.1002/pro.3420.

[17] Dasari AKR, Hung I, Gan Z, Lim KH. Two distinct aggregation pathways in transthyretin misfolding and amyloid formation. Biochimica et Biophysica Acta (BBA) – Proteins & Proteomics 2018. https://doi.org/10.1016/j.bbapap.2018.10.013. pii: S1570-9639(18)30188-2.

[18] Ankarcrona M, Winblad B, Monteiro C, Fearns C, Powers ET, Johansson J, Kelly JW. Current and future treatment of amyloid diseases. Journal of Internal Medicine 2016;280(2):177–202. https://doi.org/10.1111/joim.12506.

[19] Balch WE, Morimoto RI, Dillin A, Kelly JW. Adapting proteostasis for disease intervention. Science 2008;319(5865):916–9. https://doi.org/10.1126/science.1141448.

[20] Coelho T, Merlini G, Bulawa CE, Fleming JA, Judge DP, Kelly JW, Huertas P. Mechanism of action and clinical application of tafamidis in hereditary transthyretin amyloidosis. Neurology and Therapy 2016;5(1):1–25. https://doi.org/10.1007/s40120-016-0040-x.

[21] Eisele YS, Monteiro C, Fearns C, Encalada SE, Wiseman RL, Powers ET, Kelly JW. Targeting protein aggregation for the treatment of degenerative diseases. Nature Reviews Drug Discovery 2015;14(11):759–80. https://doi.org/10.1038/nrd4593.

[22] Schonhoft JD, Monteiro C, Plate L, Eisele YS, Kelly JM, Boland D, Kelly JW. Peptide probes detect misfolded transthyretin oligomers in plasma of hereditary amyloidosis patients. Science Translational Medicine 2017;9(407). https://doi.org/10.1126/scitranslmed.aam7621.

Summary: protein is an intelligent micelle

Leszek Konieczny[1], **Irena Roterman**[2]
[1]Chair of Medical Biochemistry, Jagiellonian University – Medical College, Krakow, Poland
[2]Department of Bioinformatics and Telemedicine, Jagiellonian University – Medical College, Krakow, Poland

Contents

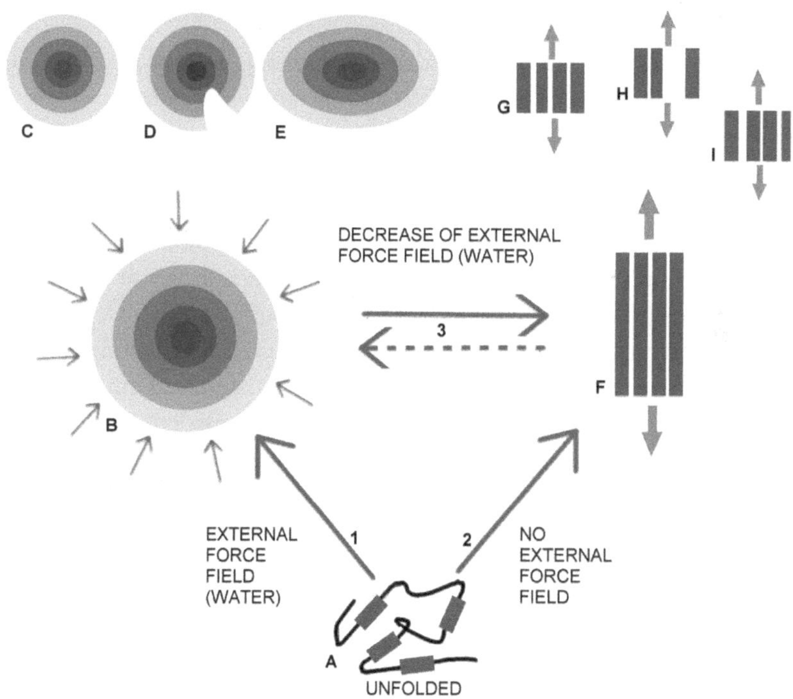

Folding or misfolding. A – unfolded form of polypeptide B – spherical micelle as the result of water environment influence (path 1), C,D,E – different forms of locally

From Globular Proteins to Amyloids
ISBN: 978-0-08-102981-7
https://doi.org/10.1016/B978-0-08-102981-7.00021-X

discordant spherical micelle, F — amyloid form with linear propagation of bands of different hydrophobicity level (low influence of water environment — path 2). Path 3 — possible transformation of globular form to amyloid form. Dashed line — path not observed however the search of reverse transformation highly expected for therapy.

Schematic depiction of alternative folding pathways, depending on the influence of the aqueous environment.

Path 1: transition between the unfolded structure (A) and an ordered form conditioned by the influence of water (micelle-like structure; B)

Path 2: weak influence of the external force field results in the folding process being dominated by the intrinsic properties of each residue, and — consequently — in linear propagation of bands characterized by variable hydrophobicity (F)

Path 3: amyloid transformation resulting from a decrease in the influence of the aqueous environment.

Path 3 (dashed line): hypothetical reverse transformation of an amyloid form into a globular form.

C, D, E — various local deformations in the structure of a micellar hydrophobic core.

G, H, I — various linear patterns depending on the sequence of the polypeptide.

This chapter discusses the potential for structural transformation represented by the dashed line (reverse path 3).

Summarizing all the collected results, we can state that the spontaneous process caused by the influence of the aqueous environment upon molecules characterized by variable hydrophobicity produces a concentration of hydrophobic residues at the center of the protein body, along with the corresponding exposure of hydrophilic residues on its surface (Chapters 1 and 2). Using antifreeze and fast-folding proteins as examples, we provide evidence of protein structures which resemble near-perfect spherical micelles. The role of a prominent hydrophobic core is particularly evident in the case of fast-folding proteins, which — when unfolded — are capable of near-instantaneous reversion to their native form. It appears that this process is driven mainly by hydrophobic interactions. The information content of such proteins may be regarded as low, given their highly ordered, symmetrical structure (Chapters 3 and 4.). Recreating such a structure following disruptions should be relatively easy — this determinism is therefore important for the proteins' functional properties (Chapter 5).

Our analysis of lysozymes, ribonuclease and other proteins shows that a hydrophilic sheath encapsulating a hydrophobic core is consistent with the theoretical distribution of hydrophobicity, and also with the structure of a spherical micelle. We may conclude that the sheath forms naturally, as a result of interaction between the polypeptide chain and the aqueous environment. With regard to single-chain enzymes, we note that they also resemble spherical micelles — at least to a substantial degree. The discussed examples show that some polypeptide chain of certain amino acid sequence are not able to generate the spherical micelle. This disability coded in amino acids sequence appears to be responsible for specific discordance. This specific discordance appears to carry information determining the biological activity of discussed protein (Chapter 6).

In the case of local deviations from the theoretical distribution of hydrophobicity, these may manifest either as deficiencies — typically corresponding to binding cavities (e.g. in the case of enzymes, as discussed above) — or local excess, which is particularly noteworthy if it occurs on the surface. Both types of deviations encode information required to attract a bind a specific ligand (in the former case) or form a complex with another protein (in the latter case).

The discordance appears to be coded in amino acid sequence.

Influence of protein on water environment

Encoding of information is also required in order to send signals out to the environment. Variable properties of the protein surface — particularly relation between polar and non-polar areas - may affect the structure of the surrounding solvent, which differs (in subtle ways) depending on what type of surface it remains in contact with. This effect can be observed on the example of antifreeze proteins, which possess a wide range of signaling capabilities — from a near-perfect micelle all the way to complex structures which include solenoid fragments, representing ordered deviations from the Gaussian distribution of hydrophobicity. The essential purpose of antifreeze proteins is to dispatch signals to the environment in such a way as to prevent formation of ice crystals. If the solenoid is capable of sending signals, why wouldn't other structures be able to do the same? (Chapter 7). Advantage coming from fuzzy oil drop model is the possibility to measure all these effects in quantitative way.

The degree of complexity increases further when we consider proteins which contain multiple discordant fragments (local excess or local deficiencies). Many large proteins deviate from the monocentric distribution of hydrophobicity because they consist of multiple domains. In such cases, however, each domain, when analyzed on its own, is typically found to adhere to the theoretical model, i.e. it resembles a spherical micelle with a prominent hydrophobic core. Assembling several domains produces a structure which – in its entirety – does not follow a Gaussian distribution of hydrophobicity. This effect is also a means of encoding information since the placement of each component domain is precisely determined (Chapter 8).

One interesting example is provided by lyase, whose complex structure appears to fulfill many conditions required for biological activity. In order to support its mechanism of action, the protein generates an internal force field. Which, when confronted with the surrounding environment, creates specific conditions facilitating the process of catalysis. The structure of this protein includes a fragment which guarantees solubility (parts of molecule accordant with assumed model), despite the presence of fragments which significantly deviate from the micelle-like conformation promoted by the surrounding water (Chapter 8). The discordance versus the idealized 3D Gauss distribution on the other hand influences the surrounded water directing the water molecules ordering in a specific form. This influence appears to be quite complicated and differentiated in lyases (Chapter 7). The force field present in protein molecule is not only influencing the water environment. It is delivering the specific force field for catalytic reaction. Solenoid with linear order of hydrophobicity distribution in lyeases is isolated from the water contact. Is seems to generate the specific local force field which probably is necessary for catalytic reaction.

Information – regardless of quantity – encoded in each protein causes fragments of the structure to deviate from the theoretical model while other fragments remain consistent with the Gaussian distribution (and therefore produce a micellar structure). It therefore seems more interesting to speculate about the structural properties of discordant fragments rather than of fragments which conform to the model. The question of how a protein chain reaches a conformation which includes discordant fragment may be addressed in two ways: either the specific sequence of amino acids directly encodes areas of discordance, or the discordance emerges as a result of information coming from an external source. In either case, we may conclude that no single, uniform and general method may be applied to all possible protein structures to produce specialized, targeted biological activity. One uniform procedure of optimization (energy minimization) seems not to be sufficient to generate the order accordant with energy minimization on one hand and local discordance on the other. The final structure seems to be the result of specific consensus between internal force field (interatomic interaction) with external force field (influence of environment) (Chapter 2).

The presented scale of structural complexity also includes molecular robots (Chapter 4). While no specific example is presented here, an analysis of the GroEl chaperonin [1] is a very big construction with highly complex structure may be found in Ref. [2]. The big amount of information necessary in such case comes from — multi-chain (21 chains) and multi-domain (14 chains — 3 domains each) construction. This "robot" is expected to act as folding chamber. The structure of GroEl — highly symmetrical in relaxed form — during performing its job looses its symmetry completely [1]. The symmetry is necessary to find the way to return to the initial relaxed form.

Some proteins — as shown in the Chapter 9 — require permanent presence of a complementary molecule which modulates the external force field. In the absence of this chaperone (which provides additional external information) the base structure reverts to an information-free state, i.e. to a ribbonlike micelle. Ribbon-like structures may, in principle, be generated for any sequence of amino acids in which strong fluctuations of hydrophobicity are confined to short fragments. Amyloid transformation may therefore be viewed as a process by which a structure which encodes information converts to an information-free form. Analysis of amyloid structures listed in PDB helps us establish specific criteria for identification of amyloids, as explained in Chapter 10.

Solenoids, where bands of high and low hydrophobicity propagate in a manner similar to amyloids, are found in biologically active proteins. In such proteins, solenoids are typically equipped with additional structural elements which counteract unchecked complexation and ensure solubility — these "stop" fragments may be analyzed in order to devise new methods of preventing formation of amyloid fibrils, as suggested in Chapter 11.

On the basis of the fuzzy oil drop model, we propose a set of criteria for identifying amyloid structures — these include the presence of a linear arrangement of alternating bands of high and low hydrophobicity, stretching along the axis of the fibril. In terms of fuzzy oil drop model parameters, amyloid structures are characterized by high values of RD (for both: T-O-R as well as for T-O-H) along with low (even negative) HvT and TvO correlation coefficients and high HvO correlation coefficients. Altogether, these values represent a specific type of discordance versus T, which may be regarded as systemic — instead of the tendency to generate a hydrophobic core, the polypeptide chain adopts an entirely different structural pattern.

If the above observations are correct, it becomes possible to identify fragments which may promote amyloid transformation in properly folded proteins. In this publication, we focus on transthyretin, which is a known amyloid precursor. Of course, the structure of transthyretin supplied by PDB is not an amyloid – in order to become an amyloid, it must undergo conformational rearrangement, which is a speculative process – however, our point is that such rearrangement is not ruled out solely by the protein's structural properties (Chapter 12.).

In the authors' opinion the presented model meshes well with the observed properties of protein folding, as well as with pathological changes which lead to misfolded proteins. Further studies of amyloidosis should, first and foremost, acknowledge the structural properties of the surrounding medium (i.e. water), which determines the properties of the external force field and provides a ubiquitous background for processes occurring in living organisms.

Influence of water environment on amyloid transformation

The view expressed in this work is that amyloidogenesis, which – as it turns out – does not always require chemical changes in the protein, results from abnormalities in protein-water interactions. The aqueous solvent enables the protein to achieve the correct fold and therefore become biologically active. Thus, analyses of misfolding phenomena should focus on the structural properties of the solvent and the external force field generated by it.

Specific facets of the problem which, in our view, merit attention include:

1. Structural properties of pure water – lack of universal water force field definition
2. Structural effects caused by addition of 0.9% of NaCl
3. Causative link between increasing concentrations of NaCl and reversible denaturation of proteins
4. Phase boundary effects – structural properties of water surfaces and their interaction with ambient air – this is related to the widely used technique of producing amyloids by shaking (which increases the phase boundary surface area)
5. Effect of SDS on the structural properties of water (as opposed to its effect on proteins)
6. Effect of DMSO on the structural properties of water (as opposed to its effect on proteins)

7. Effects caused by exposure of hydrophobic surfaces – this research is ongoing, and the presented work cites some available results [3–6].

The notion of "structural properties" refers to specific arrangement of water molecules which gives rise to a continuous force field. It would be useful to investigate potential intermolecular communication channels which rely on enforcing a certain structural order (or disorder) in the aqueous medium. This has already been attempted in the so-called iceberg model [7–9], however, in the Authors' view, a more comprehensive approach is required.

If the presented model is based on correct assumptions, it can be applied to design structures capable of arresting unrestricted propagation of amyloid fibrils. By analyzing fragments which appear to perform this function in certain active proteins, we can propose artificial "caps" – amphipathic helices – capable of binding to the tip of the fibril (with the required specificity) and exposing a hydrophilic surface, thus preventing unchecked growth [10,11], (Chapter 11). The amphipatic character of stoppers is not limited to helical forms. Any structure which satisfy the compatibility to elongated fibril on one site and introducing hydrophilic part on the opposite site may play a role of stoppers.

Regarding theoretical research (computerized simulations), the authors are interested in simulating the folding process under variable external force fields which can be modeled as dynamic changes in the structure of the encapsulating 3D Gaussian. It seems that this process may produce conditions which favor the production of spherical and/or ribbonlike micelles devoid of biological information. The experiment would also enable us to quantify the role of the environment in ensuring that proteins attain their native forms. The presented research should be therefore considered prospective in character (Chapter 9).

The influence of chemical and physical factors widely discussed in Refs. [12,13]. The results of these experiment shall be consumed not only as factors influencing structural changes in protein. They should be consumed in form of water force field construction treated as continuous medium sensitive to environmental factors (pH, presence of ions, temperature etc).

Protein is an intelligent micelle

In summary one shall conclude, that controlled local discordance (in respect to spherical micelle hydrophobicity concentration in central part of protein molecule) is aim-oriented. The degree of his discordance is highly differentiated. It encodes the specific biological activity of proteins. This local doscordance is carrying information determining the biological activity. It seems to be reached as the effect of the consensus between external force field (influence of environment) and internal force field (inter-atomic interaction).

Bi-polar molecules — as all amino acids — in water environment tend to minimize the disadvantageous entropic hydrophobic/hydrophilic effects. As long as idealized spherical micelle can be generated it appears as the native final structure. The determined neighborhood (peptide bonds) of differentiated hydrophobicity/hydrophilicity obligates to generate more or less ordered micellar forms.

In case when the idealized spherical micelle is impossible to be generated other form minimizing the unfavourable hydrophobicity-water contact the ribbon-like micelle is constructed.

The molecules which undergo the amyloid transformation appearing are acting in vivo in form of complex (called here permanent chaperones — Chapter 9). Deprivation of the permanent target with disability to generate the spherical micelle directs these proteins to generate the only possible structure minimizing the exposure of hydrophobic regions which is the ribbon-like micelle.

Despite the larger than usual size of the publication it is impossible to discuss other models of amyloidogenesis interpretation what has been the subject of many publications over the years. Readers can familiarize themselves with progress in this field by referring to comprehensive reviews such as Chiti and Dobson [14—16] and two books edited by Prusiner [17,18]. It is, however, difficult to compare the presented work with other published studies given that other authors do not acknowledge the effect of hydrophobic interactions upon amyloid transformation. The influence of water environment is widely discussed [3—6,19—39] as well as amyloids are the objects of many papers which may be recognized as complementary to the model presented in this work [40—64].

We hope that presented here discussion may introduce interpretation of amyloidogenesis from the point of view of environment influence on protein structure and misfolding in particular.

We hope also that the interpretation of proteins as intelligent (spherical) micelles and amyloids as ribbon-like micelles deprived of any form of information carried appears legimited.

References

[1] Xu Z, Horwich AL, Sigler PB. The crystal structure of the asymmetric GroEL–GroES–(ADP)7 chaperonin complex. Nature 1997;388(6644):741–50. https://doi.org/10.1038/41944.

[2] Banach M, Stąpor K, Roterman I. Chaperonin structure – the large multi-subunit protein complex. International Journal of Molecular Sciences 2009;10(3):844–61. https://doi.org/10.3390/ijms10030844.

[3] Panganiban B, Qiao B, Jiang T, DelRe C, Obadia MM, Nguyen TD, Xu T. Random heteropolymers preserve protein function in foreign environments. Science 2018; 359(6381):1239–43. https://doi.org/10.1126/science.aao0335.

[4] Lupi L, Hudait A, Peters B, Grünwald M, Gotchy Mullen R, Nguyen AH, Molinero V. Role of stacking disorder in ice nucleation. Nature 2017;551(7679): 218–22. https://doi.org/10.1038/nature24279.

[5] Schutzius TM, Jung S, Maitra T, Graeber G, Köhme M, Poulikakos D. Spontaneous droplet trampolining on rigid superhydrophobic surfaces. Nature 2015;527(7576): 82–5. https://doi.org/10.1038/nature15738.

[6] Macias-Romero C, Nahalka I, Okur HI, Roke S. Optical imaging of surface chemistry and dynamics in confinement. Science 2017;357(6353):784–8. https://doi.org/10.1126/science.aal4346.

[7] Ben-Naim A. Myths and verities in protein folding theories: from Frank and Evans iceberg-conjecture to explanation of the hydrophobic effect. The Journal of Chemical Physics 2013;139(16):165105. https://doi.org/10.1063/1.4827086.

[8] Ben-Naim A. Solvent-induced forces in protein folding reflections on the protein folding problem. Current Opinion in Colloid and Interface Science 2013;18(6): 502–9. https://doi.org/10.1016/j.cocis.2013.11.001.

[9] Naim AB. Myths and verities in protein folding theories Part II: from Kauzmann's conjecture to the dominance of the hydrophobic effect. Journal of Advances in Chemistry 2007;3(2):205–12. https://doi.org/10.24297/jac.v3i2.933.

[10] Roterman I, Banach M, Konieczny L. Propagation of fibrillar structural forms in proteins stopped by naturally occurring short polypeptide chain fragments. Pharmaceuticals 2017;10(4):89. https://doi.org/10.3390/ph10040089.

[11] Roterman I, Banach M, Konieczny L. Towards the design of anti-amyloid short peptide helices. Bioinformation 2018;14(01):001–7. https://doi.org/10.6026/97320630014001.

[12] Serpell LC. Alzheimer's amyloid fibrils: structure and assembly. Biochimica et Biophysica Acta – Molecular Basis of Disease 2000;1502(1):16–30. https://doi.org/10.1016/s0925-4439(00)00029-6.

[13] Al-Garawi ZS, McIntosh BA, Neill-Hall D, Hatimy AA, Sweet SM, Bagley MC, Serpell LC. The amyloid architecture provides a scaffold for enzyme-like catalysts. Nanoscale 2017;9(30):10773–83. https://doi.org/10.1039/c7nr02675g.

[14] Chiti F, Dobson CM. Protein misfolding, amyloid formation, and human disease: a summary of progress over the last decade. Annual Review of Biochemistry 2017; 86(1):27–68. https://doi.org/10.1146/annurev-biochem-061516-045115.

[15] Chiti F, Dobson CM. Amyloid formation by globular proteins under native conditions. Nature Chemical Biology 2009;5(1):15–22. https://doi.org/10.1038/nchembio.131.

[16] Chiti F, Dobson CM. Protein misfolding, functional amyloid, and human disease. Annual Review of Biochemistry 2006;75(1):333–66. https://doi.org/10.1146/annurev.biochem.75.101304.123901.

[17] Prusiner S. Prion diseases. In: Prusiner SB, editor. Cold spring harbor perspectives in medicine; 2017.

[18] Prusiner S. Prion biology. In: Prusiner SB, editor. Cold spring harbor perspectives in medicine; 2017.
[19] Borgia A, Borgia MB, Bugge K, Kissling VM, Heidarsson PO, Fernandes CB, Schuler B. Extreme disorder in an ultrahigh-affinity protein complex. Nature 2018; 555(7694):61–6. https://doi.org/10.1038/nature25762.
[20] Kim KH, Späh A, Pathak H, Perakis F, Mariedahl D, Amann-Winkel K, Nilsson A. Response to Comment on "Maxima in the thermodynamic response and correlation functions of deeply supercooled water. Science 2018;360(6390). https://doi.org/10.1126/science.aat1729. eaat1729.
[21] Gallo P, Stanley HE. Supercooled water reveals its secrets. Science 2017;358(6370): 1543–4. https://doi.org/10.1126/science.aar3575.
[22] Palmer JC, Martelli F, Liu Y, Car R, Panagiotopoulos AZ, Debenedetti PG. Metastable liquid–liquid transition in a molecular model of water. Nature 2014; 510(7505):385–8. https://doi.org/10.1038/nature13405.
[23] Tanaka H. A self-consistent phase diagram for supercooled water. Nature 1996; 380(6572):328–30. https://doi.org/10.1038/380328a0.
[24] Biancalana M, Makabe K, Koide S. Minimalist design of water-soluble cross-β architecture. Proceedings of the National Academy of Sciences 2010;107(8): 3469–74. https://doi.org/10.1073/pnas.0912654107.
[25] Biedermann F, Nau WM, Schneider H-J. The hydrophobic effect revisited-studies with supramolecular complexes imply high-energy water as a noncovalent driving force. Angewandte Chemie International Edition 2014;53(42):11158–71. https://doi.org/10.1002/anie.201310958.
[26] Chandler D. Interfaces and the driving force of hydrophobic assembly. Nature 2005; 437(7059):640–7. https://doi.org/10.1038/nature04162.
[27] Clary DC. Quantum dynamics in the smallest water droplet. Science 2016;351(6279): 1267–8. https://doi.org/10.1126/science.aaf3061.
[28] Corradini D, Strekalova EG, Stanley HE, Gallo P. Microscopic mechanism of protein cryopreservation in an aqueous solution with trehalose. Scientific Reports 2013;3(1). https://doi.org/10.1038/srep01218.
[29] Corradini D, Su Z, Stanley HE, Gallo P. A molecular dynamics study of the equation of state and the structure of supercooled aqueous solutions of methanol. The Journal of Chemical Physics 2012;137(18):184503. https://doi.org/10.1063/1.4767060.
[30] Corradini D, Gallo P, Buldyrev SV, Stanley HE. Fragile-to-strong crossover coupled to the liquid-liquid transition in hydrophobic solutions. Physical Review 2012;85(5). https://doi.org/10.1103/physreve.85.051503.
[31] Corradini D, Buldyrev SV, Gallo P, Stanley HE. Effect of hydrophobic solutes on the liquid-liquid critical point. Physical Review 2010;81(6). https://doi.org/10.1103/physreve.81.061504.
[32] Ding F, Dokholyan NV, Buldyrev SV, Stanley HE, Shakhnovich EI. Direct molecular dynamics observation of protein folding transition state ensemble. Biophysical Journal 2002;83(6):3525–32. https://doi.org/10.1016/s0006-3495(02)75352-6.
[33] Gallo P, Amann-Winkel K, Angell CA, Anisimov MA, Caupin F, Chakravarty C, Pettersson LGM. Water: a tale of two liquids. Chemical Reviews 2016;116(13): 7463–500. https://doi.org/10.1021/acs.chemrev.5b00750.
[34] Harpham MR, Levinger NE, Ladanyi BM. An investigation of water dynamics in binary mixtures of water and dimethyl sulfoxide†. The Journal of Physical Chemistry B 2008;112(2):283–93. https://doi.org/10.1021/jp074985j.
[35] Laage D. A molecular jump mechanism of water reorientation. Science 2006; 311(5762):832–5. https://doi.org/10.1126/science.1122154.

[36] Ma CD, Wang C, Acevedo-Vélez C, Gellman SH, Abbott NL. Modulation of hydrophobic interactions by proximally immobilized ions. Nature 2015;517(7534):347–50. https://doi.org/10.1038/nature14018.
[37] Pettitt BM, Makarov VA, Andrews BK. Protein hydration density: theory, simulations and crystallography. Current Opinion in Structural Biology 1998;8(2):218–21. https://doi.org/10.1016/s0959-440x(98)80042-0.
[38] Richardson JO, Perez C, Lobsiger S, Reid AA, Temelso B, Shields GC, Althorpe SC. Concerted hydrogen-bond breaking by quantum tunneling in the water hexamer prism. Science 2016;351(6279):1310–3. https://doi.org/10.1126/science.aae0012.
[39] Strekalova EG, Corradini D, Mazza MG, Buldyrev SV, Gallo P, Franzese G, Stanley HE. Effect of hydrophobic environments on the hypothesized liquid-liquid critical point of water. Journal of Biological Physics 2011;38(1):97–111. https://doi.org/10.1007/s10867-011-9241-9.
[40] Bruinsma IB, Bruggink KA, Kinast K, Versleijen AAM, Segers-Nolten IMJ, Subramaniam V, Verbeek MM. Inhibition of α-synuclein aggregation by small heat shock proteins. Proteins: Structure, Function, and Bioinformatics 2011;79(10):2956–67. https://doi.org/10.1002/prot.23152.
[41] Jucker M, Walker LC. Self-propagation of pathogenic protein aggregates in neurodegenerative diseases. Nature 2013;501(7465):45–51. https://doi.org/10.1038/nature12481.
[42] Hartl FU. Protein misfolding diseases. Annual Review of Biochemistry 2017;86(1):21–6. https://doi.org/10.1146/annurev-biochem-061516-044518.
[43] Sandberg A, Luheshi LM, Sollvander S, Pereira de Barros T, Macao B, Knowles TPJ, Hard T. Stabilization of neurotoxic Alzheimer amyloid- oligomers by protein engineering. Proceedings of the National Academy of Sciences 2010;107(35):15595–600. https://doi.org/10.1073/pnas.1001740107.
[44] Bouma B, Kroon-Batenburg LMJ, Wu Y-P, Brunjes B, Posthuma G, Kranenburg O, Gebbink MFBG. Glycation induces formation of amyloid cross-β structure in albumin. Journal of Biological Chemistry 2003;278(43):41810–9. https://doi.org/10.1074/jbc.m303925200.
[45] Porat Y, Abramowitz A, Gazit E. Inhibition of amyloid fibril formation by polyphenols: structural similarity and aromatic interactions as a common inhibition mechanism. Chemical Biology & Drug Design 2006;67(1):27–37. https://doi.org/10.1111/j.1747-0285.2005.00318.x.
[46] Lazo ND, Downing DT. Fibril formation by amyloid-beta proteins may involve beta-helical protofibrils. Journal of Peptide Research 1999;53(6):633–40. https://doi.org/10.1034/j.1399-3011.1999.00057.x.
[47] Ferrão-Gonzales AD, Robbs BK, Moreau VH, Ferreira A, Juliano L, Valente AP, Foguel D. Controlling β-amyloid oligomerization by the use of naphthalene sulfonates. Journal of Biological Chemistry 2005;280(41):34747–54. https://doi.org/10.1074/jbc.m501651200.
[48] Nguyen KV, Gendrault J-L, Wolff C-M. Poly-l-lysine dissolves fibrillar aggregation of the alzheimer β-amyloid peptide in vitro. Biochemical and Biophysical Research Communications 2002;291(4):764–8. https://doi.org/10.1006/bbrc.2002.6514.
[49] Liu L, Murphy RM. Kinetics of inhibition of β-amyloid aggregation by transthyretin†. Biochemistry 2006;45(51):15702–9. https://doi.org/10.1021/bi0618520.
[50] Schmittschmitt JP, Scholtz JM. The role of protein stability, solubility, and net charge in amyloid fibril formation. Protein Science 2009;12(10):2374–8. https://doi.org/10.1110/ps.03152903.
[51] Hayward S, James Milner-White E. Simulation of the β- to α-sheet transition results in a twisted sheet for antiparallel and an α-nanotube for parallel strands: implications for

amyloid formation. Proteins: Structure, Function, and Bioinformatics 2011;79(11): 3193–207. https://doi.org/10.1002/prot.23154.

[52] Baiesi M, Seno F, Trovato A. Fibril elongation mechanisms of HET-s prion-forming domain: topological evidence for growth polarity. Proteins: Structure, Function, and Bioinformatics 2011;79(11):3067–81. https://doi.org/10.1002/prot.23133.

[53] Doran TM, Kamens AJ, Byrnes NK, Nilsson BL. Role of amino acid hydrophobicity, aromaticity, and molecular volume on IAPP(20-29) amyloid self-assembly. Proteins: Structure, Function, and Bioinformatics 2012;80(4):1053–65. https://doi.org/10.1002/prot.24007.

[54] Lai Z, Colón W, Kelly JW. The acid-mediated denaturation pathway of transthyretin yields a conformational intermediate that can self-assemble into amyloid†. Biochemistry 1996;35(20):6470–82. https://doi.org/10.1021/bi952501g.

[55] Potrzebowski W, André I. Automated determination of fibrillar structures by simultaneous model building and fiber diffraction refinement. Nature Methods 2015;12(7): 679–84. https://doi.org/10.1038/nmeth.3399.

[56] Dhulesia A, Cremades N, Kumita JR, Hsu S-TD, Mossuto MF, Dumoulin M, Dobson CM. Local cooperativity in an amyloidogenic state of human lysozyme observed at atomic resolution. Journal of the American Chemical Society 2010; 132(44):15580–8. https://doi.org/10.1021/ja103524m.

[57] Wu H, Fuxreiter M. The structure and dynamics of higher-order assemblies: amyloids, signalosomes, and granules. Cell 2016;165(5):1055–66. https://doi.org/10.1016/j.cell.2016.05.004.

[58] Hughes MP, Sawaya MR, Boyer DR, Goldschmidt L, Rodriguez JA, Cascio D, Eisenberg DS. Atomic structures of low-complexity protein segments reveal kinked β sheets that assemble networks. Science 2018;359(6376):698–701. https://doi.org/10.1126/science.aan6398.

[59] Gremer L, Schölzel D, Schenk C, Reinartz E, Labahn J, Ravelli RBG, Schröder GF. Fibril structure of amyloid-β(1–42) by cryo–electron microscopy. Science 2017; 358(6359):116–9. https://doi.org/10.1126/science.aao2825.

[60] Goyal P, Krasteva PV, Van Gerven N, Gubellini F, Van den Broeck I, Troupiotis-Tsaïlaki A, Remaut H. Structural and mechanistic insights into the bacterial amyloid secretion channel CsgG. Nature 2014;516(7530):250–3. https://doi.org/10.1038/nature13768.

[61] Huang K, Maiti NC, Phillips NB, Carey PR, Weiss MA. Structure-specific effects of protein topology on cross-β assembly: studies of insulin fibrillation. Biochemistry 2006;45(34):10278–93. https://doi.org/10.1021/bi060879g.

[62] Cruz SD, Cleveland DW. Disrupted nuclear import-export in neurodegeneration. Science 2016;351(6269):125–6. https://doi.org/10.1126/science.aad9872.

[63] Tomski SJ, Murphy RM. Kinetics of aggregation of synthetic β-amyloid peptide. Archives of Biochemistry and Biophysics 1992;294(2):630–8. https://doi.org/10.1016/0003-9861(92)90735-f.

[64] Fowler DM, Koulov AV, Balch WE, Kelly JW. Functional amyloid — from bacteria to humans. Trends in Biochemical Sciences 2007;32(5):217–24. https://doi.org/10.1016/j.tibs.2007.03.003.

Index

'*Note*: Page numbers followed by "f" indicate figures and "t" indicates tables.'

FSC
www.fsc.org
MIX
Papier | Fördert gute Waldnutzung
FSC® C083411

Kolibri 360 GmbH
Ferdinand-Jühlke-Straße 7
99095 Erfurt, Deutschland
produktsicherheit@kolibri360.de